Henrietta Rix Wood
913.362.2737

Environmental Science and International Politics

REACTING CONSORTIUM PRESS

This book is a "reacting" game. Reacting games are interactive role-playing games in which you, the student, are responsible for your own learning. They are used at more than 300 colleges and universities in the United States and abroad. Reacting Consortium Press is a publishing program of the Reacting Consortium, the association of schools that use reacting games. For more information visit http://reactingconsortium.org.

Environmental Science and International Politics

ACID RAIN IN EUROPE, 1979–1989, AND CLIMATE CHANGE IN COPENHAGEN, DECEMBER 2009

DAVID E. HENDERSON AND
SUSAN K. HENDERSON

© 2018 David E. Henderson and Susan K. Henderson
All rights reserved
Set in Utopia and The Sans
by Westchester Publishing Services
Manufactured in the United States of America

The University of North Carolina Press has been a member of the Green Press Initiative since 2003.

ISBN 978-1-4696-4029-7 (pbk.: alk. paper)
ISBN 978-1-4696-4030-3 (ebook)

Cover illustration: Photo of the Earth seen from Apollo 17 (NASA/Apollo 17 crew; taken by either Harrison Schmitt or Ron Evans).

Distributed by the
University of North Carolina Press
116 South Boundary Street
Chapel Hill, NC 27514-3808
1-800-848-6224
www.uncpress.org

Contents

Note to Instructors / viii

HOW TO PLAY THESE GAMES / 1

ACID RAIN IN EUROPE, 1979–1989

Contents / 8
Figures and Tables / 11

1. HISTORICAL BACKGROUND / 14

2. THE GAME / 35

3. ROLES AND FACTIONS / 41

4. CORE TEXTS AND SUPPLEMENTAL READINGS / 44

Bibliography / 106
Acknowledgments / 109
Appendices / 110

CLIMATE CHANGE IN COPENHAGEN, DECEMBER 2009

Contents / 124
Figures and Tables / 127

1. HISTORICAL BACKGROUND / 129

2. THE GAME / 153

3. ROLES AND FACTIONS / 159

4. CORE TEXTS AND SUPPLEMENTAL READINGS / 162

Acknowledgments / 164
Appendices / 165

Environmental Science and International Politics

NOTE TO INSTRUCTORS

This text contains two reacting games that deal with the intersection of environmental science and public policy. The Acid Rain game may be played in its entirety, or specific parts can be used as separate shorter games. The Climate Change game is a separate game from the Acid Rain game. The scientific issues are different in the various games, but the policy issues are very similar. Instructors should identify exactly what sections of the text are to be used.

How to Play These Games

These two games are "reacting" games. Reacting games use complex role-playing games to teach about important moments in history. After a few preparatory lectures, the game begins, and the students are in charge; the instructor serves as adviser. The games are set in moments of heightened historical tension, and they place you, the student, in the role of a person from the period. By reading the game book and your individual role sheet, you discover objectives, potential allies, and the forces that stand between you and victory. You must then attempt to achieve victory through formal speeches, informal debate, negotiations, and conspiracy. The outcomes will sometimes differ from actual history; a postmortem session sets the record straight.

What follows is an outline of what you will encounter in reacting games and what you will be expected to do.

GAME SETUP

Your instructor will spend some time before the beginning of the game helping you to understand the historical context for the game. During the setup period, you will use several different kinds of material:

- The game book (from which you are reading now), which includes historical information, the rules and elements of the game, and essential documents.
- A role sheet (provided by your instructor), which has a short biography of the historical person you will model in the game as well as that person's ideology, objectives, responsibilities, and resources.

In addition to the game book, you may also be required to read historical documents or books written by historians. These provide additional information and arguments for use during the game.

Read all this contextual material and all these documents and sources before the game begins (or as much as possible—you can catch up once the game is under way). And, just as important, go back and reread these materials throughout the game. A second and third reading while you are in the role will deepen your understanding and alter your perspective, for ideas take on a different aspect when seen through the eyes of a partisan actor. Students who have carefully read the materials and know the rules of the game will invariably do better than those who rely on general impressions and uncertain memories.

GAME PLAY

[handwritten: Dr. M. will use Robert's Rules]

Once the game begins, class sessions are presided over by students. In most cases, a single student serves as some sort of presiding officer. The instructor then becomes the Gamemaster (GM) and takes a seat in the back of the room. Though they do not lead the class sessions, GMs may do any of the following:

- Pass notes.
- Announce important events (e.g., "Sparta is invading!"). Some of these events are the result of student actions; others are instigated by the GM.
- Redirect proceedings that have gone off track.
- Call for a recess if debates become overheated.

The student presiding officer is expected to observe basic standards of fairness, but as a fail-safe device, most reacting games employ the Podium Rule, which allows a student who has not been recognized to approach the podium and wait for a chance to speak. Once at the podium, the student has the floor and must be heard.

Role sheets contain private, secret information, which students are expected to guard. You are advised, therefore, to exercise caution when discussing your role with others. Your role sheet probably identifies likely allies, but even they may not always be trustworthy. However, keeping your own counsel or saying nothing to anyone is not an option—to achieve your objectives, you *must* speak with others. You will never muster the voting strength to prevail without allies. Collaboration and coalition building are at the heart of every game.

These discussions must lead to action, which often means proposing, debating, and passing legislation. Someone thus must be responsible for introducing the measure and explaining its particulars. And always remember that a reacting game is only a game—resistance, attack, and betrayal are not to be taken personally because game opponents are merely acting as their roles direct.

Some games feature strong alliances called *factions*. These games may seem frustrating because many factions are tight-knit groups with fixed objectives. This may make finding a persuadable ally seem impossible, but fortunately these games include roles called Indeterminates, who operate outside the established factions. Not all Indeterminates are entirely neutral; some are biased on certain issues. If you are in a faction, cultivating Indeterminates is in your interest because they can be convinced to support your position. If you are lucky enough to have drawn the role of an Indeterminate, you should be pleased—you will likely play a pivotal role in the outcome of the game.

Game Requirements

Students in reacting games practice persuasive writing, public speaking, critical thinking, teamwork, negotiation, problem solving, collaboration, adapting to changing circumstances, and working under pressure to meet deadlines. Your instructor will explain the specific requirements for your class. In general, though, a reacting game asks you to perform three distinct activities: reading and writing, public speaking and debate, and strategizing.

Reading and Writing

The standard academic work of reading and writing is carried on more purposefully in a reacting course. What you read is put to immediate use, and what you write is meant to persuade others to act the way you want them to. The reading load may have slight variations from role to role, and the writing requirement will depend on your particular course. Papers are often policy statements, but they can also be autobiographies, battle plans, spy reports, newspapers, poems, or after-game reflections. In most cases papers are posted on the class website in advance of each game session for examination by others. Papers provide the foundation for the speeches delivered in class.

The exact writing requirements depend on your instructor, but in most cases you will be writing to persuade others. Most of your writing will take the form of policy statements, but you might also write autobiographies, clandestine messages, newspapers, or after-game reflections. In most cases papers are posted on the class website for examination by others. The basic rules are these: Do not use big fonts or large margins, and do not simply repeat your position as outlined in your role sheets. You must base your arguments on historical facts as well as ideas drawn from assigned texts—and from your independent research. (Your instructor will outline the requirements for footnoting and attribution.) Be sure to consider the weaknesses in your argument and address them—if you do not, your opponents will.

Public Speaking and Debate

In the course of a game, almost everyone is expected to deliver at least one formal speech from the podium (the length of the game and the size of the class will determine the number of speeches). Debate follows a speech. The debate can be impromptu, raucous, and fast-paced, and it will result in decisions voted on by the student body. The GM may stipulate that students must deliver their papers from memory when at the podium or may insist that students wean themselves from dependency on written notes as the game progresses.

Immerse yourself in the game. Regard it as a way to escape imaginatively from your usual self and from your customary perspective as a college student in the twenty-first century. At first this may feel uncomfortable because you may be advocating ideas that are incompatible with your own beliefs. You may also need to take actions that you would find reprehensible in real life. You should remember that a reacting game is only a game—you and the other players are merely playing roles. When they offer criticisms, they are not criticizing you as a person. Similarly, you must never criticize another *person* in the game. But you will likely be obliged to criticize their *persona*. (For example, never say, "Sally's argument is ridiculous." However, do feel free to say, "Governor Winthrop's argument is ridiculous"—though you would do well to explain exactly why!) When you are spoken to by a fellow player, whether in class or out, always assume that your classmate is speaking to you in your role.

Help to create this world by avoiding the colloquialisms and familiarities of today's college life. For example, the presiding officer should never open a session with the salutation, "Hi, guys." Similarly, remember that it is inappropriate to trade on your out-of-class relationships when asking for support within the game. ("Hey, you can't vote against me. We're both on the tennis team!")

Reacting to the past seeks to approximate the complexity of the past. That is, just as some people in history were not who they seemed to be, some roles in the reacting game may include elements of conspiracy or deceit. (For example, Brutus did not announce to the Roman Senate his plans to assassinate Caesar.) If you are assigned such a role, you must make it clear to everyone that you are merely playing a role. If, however, you come to find your role and actions to be too stressful or uncomfortable, tell the GM.

Strategizing

Communication among students is an essential feature of reacting games. You will find yourself writing e-mails, texting, attending out-of-class meetings, or gathering for meals on a fairly regular basis. The purpose of frequent communication is to lay out a strategy for advancing your agenda and thwarting the agenda of your opponents. You can hatch plots to ensnare the individuals who are troubling to your cause. When communicating with a fellow student in or out of class, always assume that he or she is speaking to you in role. If instead you want to talk about the "real world," you must make that clear at the outset.

Always seek allies to back your points when you are speaking at the podium. Do your best to have at least one supporter to second your proposal, come to your defense, or admonish inattentive members of the body. Note passing and side conversations, while common occurrences, will likely spoil the effect of your speech, so you and your supporters should insist upon order before such behavior becomes too disruptive. Ask the presiding officer to assist you. Appeal to the GM as a last resort.

SKILL DEVELOPMENT

A recent Associated Press article on education and employment made the following observations:

> The world's top employers are pickier than ever. And they want to see more than good grades and the right degree. They want graduates with so-called soft skills—those who can work well in teams, write and speak with clarity, adapt quickly to changes in technology and business conditions, and interact with colleagues from different countries and cultures. . . . And companies are going to ever-greater lengths to identify the students who have the right mix of skills, by observing them in role-playing exercises to see how they handle pressure and get along with others . . . and [by] organizing contests that reveal how students solve problems and handle deadline pressure.[1]

Reacting to the past, probably better than most elements of the curriculum, provides the opportunity for developing these "soft skills." This is because you will be practicing persuasive writing, public speaking, critical thinking, problem solving, and collaboration. You will also need to adapt to changing circumstances and work under pressure.

DEALING WITH SCIENTIFIC MATERIAL IN REACTING GAMES

This game deals with both science and issues of policy, ethics, and philosophy. We believe that every educated person needs to develop the ability to interpret the barrage of technical studies that are often cited but often misrepresented in the popular media. This is necessary if one is to make informed decisions on what to eat, how to deal with personal health, and how to vote on important issues of policy related to the environment.

Primary and Secondary Sources

Scientists usually publish their findings in technical journals written primarily for specialists. Few scientists will fully understand material from outside their area of specialization. The material in these journals is usually peer reviewed, which means that other specialists have reviewed the work and judged it to be reliable. The peer review process is not perfect and errors do occur, but peer-reviewed scientific publications are the gold standard for reliable scientific information. Articles from these journals are called *primary sources* because they are written by the researchers who did the actual work.

The problem faced by nonspecialists, even those trained in science, is they often cannot read

---

1. P. Wiseman, "Employers Get Pickier," AP/AOL Finance, June 26, 2013, www.aol.com/2013/06/26/top-employers-pickier-than-ever/.

or understand the research studies. So many research articles are summarized by professional science writers, generalists who read the primary literature and explain it to a wider audience of nonspecialists. *Science* and *Nature*, two of the most prestigious primary journals in the English language, often select a few of their scientific reports and explain them in the opening pages for a wider audience. These reviews or summaries are produced by science writers. Similarly, *Science News* and *Discovery* are publications that review a wide range of primary sources and summarize them for an audience of nonspecialists.

These summaries of primary sources are referred to as *secondary sources,* meaning they were not written by the people with direct knowledge of the work. Outside their own area of specialization, virtually all scientists depend on secondary literature. You would not expect a gastroenterologist to diagnose your ear infection, and biologists probably will not really understand a paper on nuclear physics. So they depend on secondary literature.

Interpreting Science in the Popular Media

The *popular media* is generally much less reliable than the secondary literature because the reporters who write for many media outlets have no real expertise as science writers. They take what they read in the secondary literature and try to make it sound interesting—or in some cases sensational—for their readers. The result of this is a steady stream of popular media reports that often appear to contradict what you may have read just days before. Caffeine is bad for you—no, now it is good. Fat is bad—no, fat is good. Well, some kinds of fat are good.

Sadly, the popular media often pick up the results of small, preliminary studies and make gross generalizations about them. Be on the lookout for these. These stories may make good television, but they are not good science. These are the studies that are the most frustrating because they often are contradicted by another study within days.

Finally, the popular media are constantly manipulated by special interest groups. Energy companies spend millions of dollars highlighting every minor flaw in the science of climate change. The food industry sponsors studies and disseminates them through the popular media to encourage you to buy their products. Makers of medical products plant glowing reports on the latest wonder drug.

Science in This Game

This game includes references to primary sources. In some courses, instructors may ask you to read the primary sources, but you should recognize that even science faculty may struggle with articles outside their general area of expertise. The good news is that most primary articles include an *abstract* that provides the basic conclusions of the study. Also, the *conclusions* toward the end of primary sources often contain the information you really need to know. The material in the middle is often a collection of the minute details that are essential for peer review, but these are unnecessary for your purposes in the game. Consequently, if you encounter difficult primary sources, make sure you look at the abstract and the conclusions.

The game book may also include *summaries* of some key primary sources. These were written by the game's authors to provide easily accessible secondary source material. At first glance, these may contain technical material that will challenge you. They may also contain information that is unnecessary to making your argument. Even if you do not fully understand a summary, try to find information that can help you to support your argument. Graphs or tables of data may be particularly easy to access.

Finally, there are some questions you should ask about all technical reports. First, ask whether it is a primary source, a secondary source, or a popular media source. In all cases, ask yourself whether the writer has a vested interest in convincing you of something. If a university laboratory writes a report on climate change, you might ask where they got their research funding. If the study's funding came from Exxon-Mobil, you may consider its information differently than if it was funded by the U.S. Environmental Protection Agency (EPA). If the study was funded by Greenpeace, that would give the information a different spin as well. The same applies to health-related publications. Are they from a company selling a product or from a research laboratory? Who paid for the study? Apply this level of critical thinking to all your sources of information. It is important to learn to read and extract what you need from a challenging document.

Mathematical Modeling
Many scientific studies use models to understand complex systems such as climate and air pollution. Models are complex mathematical equations that attempt to include all factors involved in the system. Acid rain and other pollution in Europe has been modeled using the Regional Air Pollution Information and Simulation (RAINS) model, which uses weather patterns and chemical reactions in the air to predict the outcome of various policy changes. Climate change is studied on supercomputers using models that include effects of sunlight, pollutants, and weather. None of these models is perfect. They are tested by examining how well they reproduce past events and whether they do a good job of explaining what we know has already happened. They are trusted to predict future events within limits.

Acid Rain in Europe, 1979–1989

DAVID E. HENDERSON AND
SUSAN K. HENDERSON

Contents

1. HISTORICAL BACKGROUND / 14

    Overview / 14

    Vignette: An Evening in Geneva / 14

    Time Line / 16

    Historical and Scientific Context / 17

        Air Pollution and Acid Rain / 17

            The Problem of Acid Precipitation / 18

            Effects of Acid Precipitation on the Ecosystem / 18

            Effects of Acid Precipitation on Infrastructure / 19

            International Aspects of Acid Rain / 20

        The Science of Acid Precipitation / 20

            Acids and Bases / 20

            The pH Scale / 21

            pH and Total Acidity / 25

            Parts per Million / 25

            How Does Rain Become Acidic? / 25

            What Can Be Done about Acid Rain? / 27

        The Formation of the EU and Transnational Negotiations / 29

            Organisation for Economic Co-operation and Development / 29

            European Economic Community / 30

            UN Economic Commission for Europe / 31

            Council for Mutual Economic Assistance / 31

        Nature of International Agreements / 31

        Tools of Public Policy / 32

            Command and Control / 32

            Polluter Pays / 32

            Cap and Trade / 32

    Glossary and Guide to Abbreviations / 33

2. THE GAME / 35

    Major Issues for Debate / 35

    Framing the Argument / 36

    Rules and Procedures / 36

    Role of Money in the Game / 36

    Winning the Game / 38

Outline of the Game / 38
   Setup Sessions / 38
   Game Sessions / 39
     Game Sessions 1 and 2: 1979 Geneva / 39
     Game Sessions 3 and 4: 1985 Helsinki / 39
     Game Sessions 5 and 6: 1988 Sophia / 39
   Debriefing: Postmortem / 40
Writing Assignments for the Game / 40
Counterfactual Aspects of the Game / 40

3. ROLES AND FACTIONS / 41

Public Knowledge about the Participating Countries in 1979 / 41
Faction 1: Major Industrialized Nations Opposed to a Strong Treaty / 41
Faction 2: Countries Seeking a Strong Treaty / 42
Faction 3: Eastern European Nations / 42
Faction 4: Less Developed Nations / 43

4. CORE TEXTS AND SUPPLEMENTAL READINGS / 44

Geneva Conference News / 44
   Norway / 46
Geneva Conference: Scientific Reports / 49
   Chemical Analysis / 50
   Biological Changes / 51
   Additional Reference Material / 52
Helsinki Conference News / 52
Helsinki Conference: Country-Specific Scientific Research / 63
   Sweden / 63
     Effects on Lakes / 64
     Soil Acidification / 64
   Ireland / 66
   Hungary / 67
   Austria / 68
   West Germany / 69
   Norway and Great Britain / 72
   France / 79
   Italy / 79
   Eastern Europe / 81
Sophia Conference News / 83

    Sophia Conference: Technical Background / 87
        The Science of Ozone and Smog / 87
            Health Effects of Ozone / 89
            Reduction of Nitrogen Oxide Emissions / 91
            The Internal Combustion Engine / 93
            Combustion Products and Engine Parameters / 94
            Catalysts and the Catalytic Converter / 95
            Units of Concentration—ppm, ppb, μg/dL, and μg/m³ / 96
        Lead Pollution and NO_x / 97
            Leaded Gasoline / 98
            Lead in the Environment / 100
            Lead and Health / 101
            Readings on Lead / 103
    Optional Texts on Ecology / 103
    Websites with Environmental Data / 104

Bibliography / 106

Acknowledgments / 109

Appendix 1. Introduction to Environmental Philosophy / 110

Appendix 2. Introduction to Environmental Economics / 116

Appendix 3. Using Numbers to Make Arguments / 119

Appendix 4. Study Questions for Reading Assignments / 121
    Questions on Environmental Philosophy / 121
    Questions on Lovelock's Gaia Hypothesis / 121

Figures and Tables

FIGURES

1. Effect of pH on Various Species in Lakes / 19

2. The pH Scale / 23

3. Effect of pH on Percentage of Fishless Lakes in Norway / 49

4. Sulfate Concentration and Percentage of Fishless Lakes in Norway / 50

5. The pH History in Round Loch of Glenhead Using a Logarithmic (pH) Scale / 60

6. The pH History in Round Loch of Glenhead Using a Linear Concentration Scale (Micromolar) / 60

7. Long Time Scale pH Changes in Lake Gardsohn / 64

8. Recent Changes in pH in Lake Gardsohn / 64

9. Snow and Rain Sampling Sites in Austria / 68

10. Snow Data as a Function of Altitude / 69

11. The pH Data for Sampling Sites near the Austrian Border / 70

12. Sulfate Concentration in Sampling Sites near the Austrian Border / 71

13. Sulfur Dioxide Concentrations in Great Britain / 75

14. The pH of Rain in Great Britain, 1978–1980 / 75

15. Acidification of Lakes in Norway / 78

16. Acidification of Lakes in Sweden / 79

17. Seasonal Variation in Acid Rain in Italy / 80

18. Diagram of the Processes Involved in the Formation of Ozone and Smog / 88

19. Ozone Data for Three Monitoring Stations / 89

20. Relationship between Ozone Concentration and Daily Death Rates in Twenty-Three European Cities / 90

21. Diagram of a Four-Cycle Internal Combustion Engine / 93

22. Tuning a Carburetor for Optimum Emissions and Performance / 95

23. The Catalytic Converter / 96

24. World Lead Production during the Past 5,500 Years / 100

25. Example of Game Simulator Tool / 105

TABLES

1. International Sulfur Transport / 21

2. National Sulfur Emissions, Population, Total Energy Use, and GDP / 22

3. Example of Costs of Emission Reductions Provided to Each Country / 37

4. Victory Point Scenarios for Money Spent on Treaty / 38

5. Cost of Nuclear and Alternative Energy Sources / 48

6. Annual Operating Cost for 1 Gigajoule of Energy in the United Kingdom, 1974 / 49

7. Sources of Soil Acidification by Soil Type / 65

8. Deposition of Sulfur in Three Countries / 66

9. Concentration of Pollutants Measured in Air Samples / 67

10. Wet and Dry Deposition of Sulfate and Nitrate / 68

11. Chemical Properties and Fish Population in Southern Norway / 73

12. Change in Runoff Ions from 1892 to 1976 / 81

13. Total Deposition of Sulfate and Nitrate in Four Basins / 81

14. Comparison of Streams in Four Basins / 82

15. NO$_x$ Pollution for 1985–1986 / 92

16. Comparison of Prehistoric and Modern Lead Exposure per Person / 101

17. Production and Emission of Lead / 102

18. Lead Ingestion in the United States in 1970s / 102

19. Cost Comparison for Electric Lights over Thirty Years / 120

1

Historical Background

OVERVIEW

This is a reacting game. Reacting games use complex role-play to teach about important moments in history. This game is set in a series of conferences sponsored by the United Nations (UN) that began in Geneva, Switzerland, in 1979 and continued for over a decade. The goal is to negotiate a treaty governing air pollutants transported across international borders within Europe. The game includes negotiations that followed in Helsinki, Finland, in 1984 and in Sophia, Bulgaria, in 1987. The result of these negotiations was the Long Range Transport Air Pollution Treaty, which continues in Europe to the present day.

The events that unfold in this game occur against the backdrop of larger negotiations within Europe on forming a European Economic Community (the EEC). The full details of these changes will become apparent as the game proceeds. But the fact that nations are being asked to give up some control over their energy and transportation sectors to support the general welfare of the region marks a turning point in the national sovereignty of these nations. At the time the negotiations begin, the idea of the European Union (EU) is just that, an idea in progress. National borders are still patrolled with checkpoints, and each nation has its own currency. In fact, air pollution is about the only thing that moves freely between the nations and across the Iron Curtain. The issue of transnational pollution was possibly the first area where national sovereignty was sacrificed for the common European good.

VIGNETTE: AN EVENING IN GENEVA

It has been a long day. This morning you met with the prime minister to go over your final instructions for the conference in Geneva and to plan strategy. After a quick lunch you hopped on a two-hour flight to Geneva and eventually got into your hotel room. You had dinner in your room while you read through the briefing papers one more time. Now your last task for the day is the opening reception for the conference that begins tomorrow. You would rather stay in your room, but you are expected to attend for reasons of protocol. The reception will also be a good chance to make a few useful contacts. You know many of the people who will be attending from the two years of negotiations that have led up to this meeting. You change out of your comfortable clothes into an outfit suitably professional and wearily head out to the reception.

In the ballroom there is a crush of people eating, drinking, and greeting each other heartily. Every possible nongovernmental organization (NGO) and major industry with a stake in the talks has set up shop. The industrial representatives have set up booths with bars and buffets and glamorous industrial representatives, and their hired models are trying to get the attention of the attendees. You stop by the area for the major electric utilities. You know most everyone there. They have been visiting your office for months as the preliminary negotiations developed. You grab a glass of wine from their bar and chat with them. They remind you yet again of the enormous cost of retrofitting the coal power plants that produce most of the electricity in your country to eliminate air pollution. All you can do at this point is assure them that you have read their briefs and the prime minister has made it clear that you must not allow them to be bankrupted in the negotiations to come.

As you move around the ballroom, you encounter an old friend who is representing one of the Scandinavian countries. She has been one of the driving forces that brought this conference into being. She has been harping on the problems her country faces with acid rain for almost a decade now. You tried to ignore her arguments for years, but the scientists have now backed up her assertions with solid data, and there is no getting around the fact that something must be done. She has an annoying habit of carrying a wallet full of photos with her. Most people carry pictures of their families, but she has photos of dead lakes and forests. Much as you like her, you are tired of her constant tirades. But you have to give her credit because her persistence is going to pay off. Changes are coming, and it is a foregone conclusion that this conference will make history in the environmental movement. Some kind of agreement is certain, and not everyone will be happy with it. You wish her a good evening and move on around the room.

You are suddenly accosted by someone from one of the more aggressive NGOs. They want to make the NGO's case to you personally. Your office has received correspondence and white papers from them many times in the past year or two. They are pushing for the complete elimination of coal power plants, especially the low-quality coal that is so common in much of Europe. You are diplomatic—that is the key function of your job. However, your country could never to agree to such a radical step. The only alternative to coal for some nations is nuclear power, which has its own problems. So you politely move on, stopping at another booth to refresh your glass of wine and pick up a handful of nuts from the buffet. The crush of people and the long day are starting to take their toll on you.

As you continue to circulate in the crowd, you are again surrounded by a well-organized group. These are members of the Green Party. They are the most aggressively anti-nuclear NGO at the conference. They tend to travel in packs and to make their points quite forcefully. They predict that eventually there will be a nuclear accident that will cause lots of death and illness, making whole regions around the accident uninhabitable. They also press you on the problem of disposal of nuclear waste. You are not unsympathetic to their positions, and the fact that the power industry in so many countries is based on fossil fuel means that there are strong economic forces against nuclear power in your country.

Another flashy booth is being hosted by Mercedes-Benz. They have a small string quartet playing and some comfortable seating near their booth, so you take advantage of this to sit for a few minutes. The music is relaxing, and you hope for a moment's peace. Sadly, as soon as you sit down, two of the company representatives come to sit next to you, and they proceed to give you their pitch. All the auto companies are worried about pollution controls on cars. They have seen what has happened to cars being sold in America and the extent to which all the anti-pollution devices reduce performance and add to the cost of production. They have specific figures that they "share" with you as you talk. One of them offers to let you do a test drive in a German Mercedes and a test drive in one bound for California. It is pointed out that the one bound for California suffers from a reduction in gas mileage. This is no surprise; you are well aware of this fact from your own country's auto industry. Pollution control can be costly in many ways. You can feel a headache coming on, and your feet are complaining bitterly, so it's time to thank them for their offer and move on. Besides you don't really have time for test drives.

Next, you come to a display with lots of poster-size photos of the Black Forest. It is not clear who owns this booth; it must be some NGO. But it catches your attention—you have fond memories of vacations in the Black Forest with your family as a child. You even recognize some of the places in the photos. You have been there. But when you look closely at the photos, the changes from your memories are quite frightening. The tops of most of the trees are completely dead. Many whole trees are dead, and there are no young trees growing to replace them. The Black Forest looks like it is dying. This makes you sad.

As you approach the door, headed back to your room and your bed, you encounter one last old friend. He is from Eastern Europe. You greet him warmly if a bit wearily. Mercifully, he does not want to sell you on any position. He is just celebrating the fact that he is part of this meeting. It is a real breakthrough for his nation to be invited to participate in a major European conference of this type. He thinks it is a sign of the growing détente with the West. He celebrates this and drags you over to the nearest bar for a quick toast with a glass of Russian vodka. Before he can refill your glass for a second toast, you claim exhaustion and head back to your room.

Tomorrow will be the first of many long days and nights to get this air pollution agreement hammered out. Sleep sounds good, but it does not come quickly. Your mind is still in full gear thinking about the whole environmental movement and the changes it is bringing. Your briefs are full of cost-benefit ratios and gross domestic product (GDP) numbers. There are certainly practical reasons to protect the environment. But are there deeper reasons? Is there something fundamental about the ecosystem that demands your attention? You caught sight of James Lovelock across the ballroom. You remember his controversial Gaia hypothesis that the entire earth represents a complex living organism with biological and geological processes working together to keep the environment stable and comfortable. As you drift off to sleep, you have a dream of the earth calling out to you with a smoky, hoarse voice and a cough, asking for your help. What a nightmare to end your day. And with that you slip into a deep sleep.

TIME LINE

The Scandinavian countries first focused attention on the growing acidification of their lakes and forests in the early 1970s. The vast majority of their air pollution comes from other nations. Thus, the only way they could solve their problem was to convince other nations to reduce their pollution. The clearest statement of the responsibility of nations for the impact their pollution has on other nations is found in Declaration 21 of the 1972 United Nations Conference on the Human Environment.

> States have, in accordance with the Charter of the United Nations and the principles of international law, the sovereign right to exploit their own resources pursuant to their own environmental policies, and the responsibility to ensure that activities within their jurisdiction or control do not cause damage to the environment of

other states or of areas beyond the limits of national jurisdiction.

Declaration 21 provides the legal framework for efforts to deal with transnational pollution. The Geneva conference was the first step in what would hopefully become an ongoing process of transnational pollution control.[1] These efforts were accelerated in 1975 when Leonid Brezhnev, the General Secretary of the Soviet Union, speaking at the first meeting of the Conference for Security and Cooperation in Europe (CSCE) in Helsinki, called for multinational negotiations on issues of energy, weapons, and the environment. The European nations decided to focus on environmental issues first, believing this to be the easiest area for negotiation. This decision led to the 1979 Geneva conference, which is the opening setting for this game.

HISTORICAL AND SCIENTIFIC CONTEXT

Air Pollution and Acid Rain

Air pollution is a problem that affects everyone. There are many types of pollutants and many ways to classify them.

One approach is to consider the distance over which the consequences of a pollutant are felt. When this approach is taken, we find that some forms of pollution have only local effects. Examples of this type of pollution are cigarette smoke and the carbon monoxide emitted from your car. These substances affect mostly individuals in the immediate surroundings. It is possible, at least in theory, to take direct action at the personal level to modify the behavior of the polluter.

Pollution that has the widest scope is that which affects the entire global ecology. Some examples are the emissions of chlorofluorocarbons (CFs) used in refrigerators, air conditioners, and plastic foam and carbon dioxide (CO_2). Both of these have the potential to affect the overall climate of the Earth and to cause the extinction of many species. These forms of pollution require changes in public policy by all nations if we are to minimize the damage they can cause. Either society must adapt to using less energy or must switch from carbon-based energy to alternative forms of energy (such as sun, wind, nuclear, or hydroelectric).

Acid rain lies between these extremes because it is a regional problem. We will use the simple term *acid rain* to refer to all forms of acid precipitation, including rain, snow, fog, acidic dust, and gases. The acid emissions of the United States affect Canada and the countries of Scandinavia. Scandinavia is also severely affected by pollution from Britain, France, and the Federal Republic of Germany (West Germany). The Eastern European countries, which are some of the worst polluters, suffer from their own pollution, and some is carried to surrounding countries as well. Therefore, acid rain is a pollution problem of international but not global scope.

This game will follow the efforts of the nations of Europe to forge a treaty to reduce the transnational effects of acid pollution. The time covered by the game is a twenty-year period, beginning in 1979 with a conference in Geneva. The game ends in 1989 with efforts to negotiate a multipollutant compact. Other events during this period that were significant to this process were the dissolution of the Soviet Union and the emergence of democratic institutions in many of the countries of Eastern Europe. There were also radical changes in the political landscape of Britain, West Germany, and the United States during this period. Finally, and most significantly, this period encompasses the formation of the EU and the decision by the nations of Europe to surrender some aspects of their sovereignty to the larger EU parliament.

To achieve pollution control in Europe, it is first necessary to understand the basic science of acid pollution and its effects on the ecosystem. Without this, it is impossible to build a consensus for action.

---

1. D. A. O'Sullivan, "European Concern about Acid Rain Is Growing," *Chemical and Engineering News* 63 (1985): 12–19.

Second, different understandings of humans' relationship with the earth (environmental philosophy) lead to different approaches to the problem. Third, the economic aspects of the problem must be understood. Environmental economics deals with the determination of the costs and benefits of specific environmental policies. The competing economic systems of the various nations of Europe and the structure of their energy economies will figure as important issues as the game develops. Finally, there is the public perception of the problem and the ability to organize and influence public policy. The years from 1970 to 1989, which represent the timeline for this game, were a period of increasing public interest in environmental protection.

The Problem of Acid Precipitation

Acidic precipitation affects both living and nonliving things, such as statues, stonework, bridges, roads, and works of art. Precipitation includes any form of water from the atmosphere, including rain, snow, fog, and even the settling of fine droplets of acidic particles. All forms of acid precipitation can cause damage.

One way to report the acidity of precipitation is its pH. The pH is a number that expresses the concentration of acidic particles in a solution. The lower the pH, the more acidity is present. The average pH of rain in the polluted regions is in the range of 4.0 to 5.0. Precipitation close to coal power plants may have pH values as low as 1.5. Many parts of Eastern Europe have rain with pH values in the range of 2.0 to 3.0, which is highly acidic and very damaging.

Effects of Acid Precipitation on the Ecosystem

The most serious effects of acid precipitation are on living things. It can kill or injure aquatic life, plants, and fish. It can affect entire populations by disrupting their food sources or reproduction. Damage to one species can lead to the disruption of an entire community of organisms that form a food web. A food web describes the way each species eats another. For example, if water pollution kills the tadpoles, there will be no frogs. Without frogs, the birds and animals that eat frogs will starve. Also, the insects that would have been eaten by the frogs can damage plant life that is food for rabbits. Ultimately, a small change in one species can affect entire ecosystems, either directly or indirectly.

The impact of acid rain is greatest in "sensitive areas" where the underlying rock does not contain significant amounts of limestone. Limestone is a natural rock that can neutralize acidity. It is composed of calcium and magnesium carbonates. These are called bases and can neutralize the acidity of rain. In areas of granite rock, which does not contain these carbonates, the acid rain is not neutralized, so lakes and streams become acidified more quickly. Northern Europe, southeastern Canada, the northeastern United States, and the Rocky Mountains are all "sensitive" to acidification. Thousands of lakes and tens of thousands of miles of streams have become acidified. This acidification kills large numbers of fish. In addition, acidification can also kill plants and animals in the water that serve as food for fish.

Figure 1 summarizes what happens to fish as the pH decreases. As the pH drops to lower numbers, various species of fish die out, and the lake becomes limited to a few organisms. One way to identify an acidified lake is that the water is crystal clear, due to the lack of microscopic organisms called plankton.

Acid precipitation causes a decline in species diversity. Different species of plants and animals have different tolerance ranges for acidity as shown in Figure 1. Species that are beyond their acid tolerance levels decline or disappear. Those species that tolerate higher acid levels grow and dominate. Typically there are far fewer acid-tolerant species, so community diversity declines with acidification. Studies in ecology have demonstrated that the more diverse a community, the more stable it is. Decreased diversity is usually seen below pH 6.0.

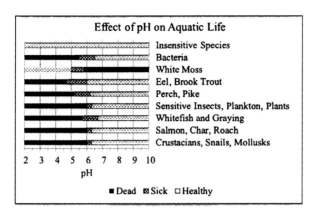

FIGURE 1 Effect of pH on various species in lakes. (Data from Park, 1987, p. 78.)

The most dramatic symptom of acidification is the massive fish deaths reported in Scandinavia, southeastern Canada, and the northeastern United States. Some of these fish deaths are caused directly by high acidity, but the majority of fish may have died from exposure to heavy metals. This is related to high acidity, as shown by Gene Likens in 1985, who found that acidified lakes had abnormally high levels of dissolved aluminum, cadmium, mercury, and other metals.[2] Acid dissolves heavy metals in the soil (that is, in the bottom sediment or surrounding soil) and then carries these metals into streams, lakes, and groundwater. Many heavy metals are highly toxic and are passed along the aquatic food chain, becoming more toxic as they accumulate. Acidity and metal toxicity have a synergistic effect, in which the several metals together produce an effect greater than the sum of their individual effects.

Acid deposited on trees filters nutrients out of the leaves, causing a decline in the trees' health. However, the most important effect of acid deposition comes from the indirect effect of the acidification of soils. The pH of a soil influences the uptake of soil nutrients by plants. Acidification decreases the availability of most nutrients, making trees more vulnerable to drought, disease, and pests. Acidification of soil also mobilizes heavy metals (aluminum, lead, mercury, and cadmium), which damages tree roots, making them more susceptible to attacks by drought, disease, insects, or fungi and mosses that thrive under acidic conditions.[3] Acidity, nutrient loss, and heavy metals also lower the germination rates of tree seeds and growth rates of seedlings.[4] World attention has focused on the damage to fir and spruce trees in the Black Forest in West Germany, but forest damage is now widespread throughout northern Europe.

Finally, acid precipitation disrupts the normal nitrogen cycle. Nitrogen-fixing bacteria are necessary to convert nitrogen in the air into forms that are used by plants. Nitrogen-fixing bacteria are eliminated at low pH levels; signs of nitrogen-cycle disruption are seen in lakes with waters below pH 5.5. The bacteria and fungi that decompose dead organic matter also die at low pH levels. These organisms are important for recycling the nutrients in dead organisms back into the ecosystem. Slowing the decomposition rate of organic matter traps nutrients inside dead organisms, which then are buried by sediment and permanently removed from the ecosystem.

Effects of Acid Precipitation on Infrastructure

Acid precipitation damages buildings, roads, bridges, works of art, and monuments. Acids attack stone, concrete, and metal. Acids dissolve marble, which is made of calcium carbonate, and other types of rock. The great marble remains of ancient Rome and Athens have been more seriously damaged by the past twenty years of air pollution than by contact with air and rain for the previous 2,000 years. Nor is the damage just aesthetic—the repair and replacement of buildings, bridges, and

---

2. G. E. Likens, "An Experimental Approach to the Study of Ecosystems," *Journal of Ecology* 73 (1985): 381–96.

3. B. Ulrich, R. Mayer, and P. K. Khanna, "Chemical Changes Due to Acid Precipitation in a Loess-Derived Soil in Central Europe," *Soil Science* 130 (1980): 193–99.

4. G. Abrahamsen, A. O. Stuanes, and B. Tveite, "Effects of Long Range Transported Air Pollutants in Scandinavia," *Water Quality Bulletin* 8 (1983): 89–95, 109.

other works costs all of us money. Estimates of the annual cost of the destruction due to acid precipitation are substantial, running to many billions of dollars each year.

International Aspects of Acid Rain

The Scandinavian countries were the first to notice and publicize the damage being done by acid rain. There are some indications of similar problems in the Black Forest of West Germany, but this is not yet believed to be severe. The damage is the result of both acid rain and ozone pollution linked to nitrogen oxides. An even larger problem was revealed by the opening of Eastern Europe to Western observers. The results of the total disregard of air pollution in Eastern Europe were shocking when finally revealed.

The greatest damage caused by air pollution in Eastern Europe can be seen in the "Black Triangle" on the border of the Democratic Republic of Germany (East Germany), Poland, and Czechoslovakia. Lignite, known as "brown coal," has been burned for industrial production, electricity generation, and heat. Brown coal is a very low energy yielding coal that is high in sulfur, up to 15 percent. Levels of sulfur dioxide in the air in Prague have reached values 400 times greater than the maximum acceptable value for air in the United States. As a result, acid rain and smog have completely destroyed large areas of forest. People living in the area also suffer from high rates of respiratory disease as a direct result of the polluted air they breathe. Many areas are ecological disaster areas.

The reason acid pollution must be dealt with as an international rather than a national problem becomes evident when one examines the data in Table 1, which shows the preliminary results of the Evaluation of Long Range Transport of Air Pollution in Europe (EMEP). This was the first evidence that pollution from some countries was transported to others. It also allowed the countries of Europe to be divided into those who receive more pollution from others than they produce and the countries that receive mostly their own pollution but export to others.

Note that some countries like the United Kingdom and Italy receive most of their sulfur emissions as local acid pollution; others, most notably Norway, Austria, and the Netherlands, receive most of their acid pollution from other countries' emissions. This produces a real dividing line in both national and international pollution policy. The former group would presumably profit considerably from local emission controls because it would reduce environmental damage to their citizens; the latter group has little to gain by implementing expensive air pollution regulations. On the other hand, the nations who import a majority of their pollution have a strong incentive to get other nations to control their air pollution.

It might seem reasonable to assume that countries like the United Kingdom and Italy would be cooperative on reducing emissions in their own national interest, but the opposite was in fact the case. Neither the United Kingdom nor Italy had strong environmental movements in 1984, and there was little awareness of the damage caused by air pollution within those countries.

The economic strength of each country and its presumed ability to invest in pollution control are related to its per capita gross domestic product (GDP). The values in Table 2 are the estimated values for 1985 based on assumptions for growth and inflation and converted to U.S. dollars. The values for the Eastern European nations are very crude estimates based on relative purchasing power of the citizens as compared with those in West Germany.

The Science of Acid Precipitation
Acids and Bases

Acids taste sour because they react with specific sites on your tongue to produce the feeling that we all learn to call sour. Some chemicals can destroy (neutralize) acid; these are called bases. *Bases,* such as the hydroxide ion OH^-, are most commonly discussed as substances that can react with the

TABLE 1 International sulfur transport

Country	SO₂ Emissions (million metric tons/year)	Percentage of SO₂ Deposition		
		Indigenous	Foreign	Unknown
Mostly local pollution				
U.K.	4.70	79	12	9
Italy	3.33	70	22	8
Ireland	0.17	28	32	40
USSR	25.50	53	32	15
East Germany	4.00	65	32	3
France	3.40	52	34	14
Mostly imported pollution				
West Germany	3.50	48	45	7
Poland	3.00	42	52	6
Belgium	0.73	41	53	6
Denmark	0.46	36	54	10
Hungary	1.50	42	54	4
Finland	0.54	26	55	19
Czechoslovakia	3.37	37	56	7
Sweden	0.25	18	58	24
Norway	0.15	8	63	29
Netherlands	0.51	23	71	6
Austria	0.43	15	76	9

Sources: Wetstone and Rosencranz, 1983, 16–20; data taken from EEC EMEP/MS—W Report, January 1981. For purposes of the game these data are assumed to be available before final publication to the EEC and those attending the Geneva Conference.

hydronium ion to produce a molecule of the solvent water:

$$H_3O^+ + OH^- \rightarrow 2H_2O$$

When bases are added to water, the solutions are said to be alkaline or basic. Acids and bases are opposites, and solutions that contain more H_3O^+ than OH^- are referred to as being acidic. Solutions that contain more OH^- than H_3O^+ are basic. Solutions that contain equal amounts of H_3O^+ and OH^- are said to be neutral.

The pH Scale

The term *pH* is used to describe the relative acidity or alkalinity of aqueous solutions. Before we discuss acid rain, it is first necessary to define acids and understand the way scientists quantify them.

TABLE 2 National sulfur emissions, population, total energy use, and GDP

Country	SO_2 Emissions in Million Metric Tons per Year[a]	Population (millions)[a]	Total Energy in 10^{15} BTU per Year[a]	GDP Billion US$, Projected (1985)[b]	SO_2 per Million People	SO_2 per 10^{15} BTU per Year	SO_2 per Billion $GDP
					\multicolumn{3}{c}{SO_2 in Thousands of Metric Tons/Year Divided by Data in Other Columns}
Italy	3.33	56.2	5.22	393	59.3	637.9	8.5
Norway	0.15	4.0	0.69	58	37.5	217.4	2.6
Ireland	0.17	3.1	0.36	18	54.8	472.2	9.4
Sweden	0.25	8.1	1.44	123	30.9	173.6	2.0
Austria	0.43	7.6	0.89	80	56.6	483.1	5.4
Denmark	0.46	5.1	0.83	81	90.2	554.2	5.7
Netherlands	0.51	13.7	2.25	176	37.2	226.7	2.9
Finland	0.54	4.7	0.69	46	114.9	782.6	11.7
Belgium	0.73	9.8	1.80	137	74.5	405.6	5.3
Hungary[c]	1.50	10.5	1.05	45	142.9	1,428.6	33.3
Poland[c]	3.00	34.0	5.58	80	88.2	537.6	37.5
Czechoslovakia[c]	3.37	14.6	3.22	50	230.8	1,046.6	67.4
France	3.40	52.6	6.64	770	64.6	512.0	4.4
West Germany	3.50	61.8	10.72	873	56.6	326.5	4.0
East Germany	4.00	16.9	3.44	40	236.7	1,162.8	100.0
U.K.	4.70	56.0	8.44	452	83.9	556.9	10.4
USSR	25.50	268.2	41.44	—	95.1	615.3	—

Note: Estimates for purpose of game where data are unavailable. BTU = British thermal unit; GDP = gross domestic product.
[a] *Source*: "Co-operative Programme for Monitoring and Evaluation of the Long-Range Transmission of Air-Pollution in Europe (EMEP)," EMEP/MSC-W Report 1/84 (1984), cited in Wetstone and Rosencranz, 1983, pp. 16–17.
[b] *Source*: World Bank, 2007.
[c] *Source*: Data derived by extrapolation from data in Andonoiva, 2004.

One of the simplest ways to describe acids is that when you put them in water, the water tastes sour. This is a very old way of classifying chemicals, and our modern understanding of acids is much more sophisticated. Acids, such as hydrogen chloride (HCl), react with water (H_2O) to produce hydronium ions, H_3O^+ (often abbreviated as the hydrogen ion, H^+) and chloride ions (Cl^-):

$$HCl + H_2O \rightarrow H_3O^+ + Cl^- \text{ or } HCl + H_2O \rightarrow H^+ + Cl^-$$

Chemical equations and formulas use the abbreviation letter of each atom with a subscript to indicate the number of each type of atom. *Ions* are atoms or groups of atoms that either have extra electrons, leading to negative charges, or have lost electrons, leading to positive charges. The charges on ions are indicated with superscripts.

The *concentration* of hydronium ions in aqueous solutions can vary over a very large range, about fourteen orders of magnitude. This enormous range from 1 to 0.00000000000001 (1×10^{-14}) moles per liter is challenging to discuss without resorting to numbers that are hard to read and understand. Chemists have developed a system to simplify the discussion of acidic and basic solutions. This relies on defining a scale called the *pH scale*. The pH scale is logarithmic, and pH is formally defined as the negative logarithm of the hydrogen ion concentration. If we write this as an equation we have,

$$pH = -\log [H^+]$$

The square brackets around H^+ simply mean concentration.

For water solutions, pH varies from 0 to 14.0, which is much easier to talk about than numbers ranging from 1.0 to 1.0×10^{-14}. Because the pH scale is based on powers of ten, a change of 1.0 pH—for example, from pH 4.0 to 3.0—corresponds to a change in hydrogen ion concentration of a factor of ten. Thus, a solution of pH 3.0 has ten times more hydrogen ions than one of pH 4.0.

Figure 2 shows the pH of many common substances. When the pH is <7.0, the solution is *acidic,* and when the pH is >7.0, the solution is *basic.* Solutions of pH=7.0 are *neutral.* A low pH of 1.0 corresponds to a relatively large concentration of acid: 0.10 moles per liter. A number of common substances are quite acidic, as shown in Figure 2.

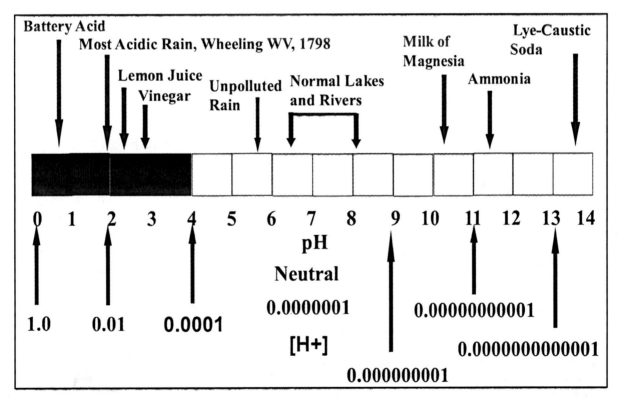

FIGURE 2 The pH scale.

Strong acids such as sulfuric acid (H_2SO_4), nitric acid (HNO_3), and hydrochloric acid (HCl) are corrosive poisons, and they cause immediate damage to living tissue at high concentrations.

Completely pure water prepared in a vacuum is neutral, and it has a pH of 7.0 at 25°C. This corresponds to a concentration of 1.0×10^{-7} moles/liter (0.0000001) of hydrogen ions. Blood is very close to neutral, with normal pH values ranging from 7.2 to 7.4.

When the concentration of hydrogen ions decreases to a concentration lower than that of pure water, the solution is said to be basic, or *alkaline*. A number of basic substances are also indicated in Figure 2. Strongly basic substances are also corrosive poisons. For example, sodium hydroxide (NaOH), which is found in lye and caustic soda, is a common strong base that causes severe burns. Strong bases are often more hazardous to handle than strong acids because they dissolve body tissues like skin and eyes. (You should always use eye protection when handling acids and bases.)

Acids and bases are often divided into two groups, strong and weak. We will limit our discussion here to acids because we are concerned with the problem of acid rain. The bases we will consider are those found in rocks and soils, which we refer to as *carbonates*.

The strong acids—hydrochloric, nitric, and sulfuric acids, and a few others—all produce a mole[5] of hydrogen ions for every mole of acid that is placed in water:

$$HNO_3 + H_2O \rightarrow H_3O^+ + NO_3^-$$

---

5. A *mole* refers to 6.02×10^{23} molecules, atoms, or ions. The number turns out to be a convenient way to group atoms or molecules, which makes the arithmetic of chemistry easier. If you know the atomic weight of a chemical and collect that many grams of the substance, you find you always have the same number of atoms or molecules, 6.02×10^{23}.

Sulfuric acid produces almost 2 moles of hydrogen ions for each 1 mole of acid. These reactions are said to go to completion, leaving no unreacted starting materials (reactants) in solution:

$$H_2SO_4 + 2H_2O \rightarrow 2H_3O^+ + SO_4^{2-}$$

Weak acids, on the other hand, do not entirely go to completion, so there is always a large amount of the starting materials (reactants) left over. What this means is that this type of acid produces only a few hydrogen ions when added to water. For acetic acid (abbreviated as HOAc for simplicity), the acid found in vinegar, only about 0.03 moles of hydrogen ions are produced when 1 mole of the acid is added to a liter of solution.

$$HOAc + H_2O \rightleftharpoons H_3O^+ + OAc^-$$

This reaction produces only about 3 percent products and leaves 97 percent of the original acetic acid unreacted. Thus, the weak acids are much less effective sources of hydrogen ions and are less effective in reducing the pH of water than the strong acids.

Another weak acid is carbonic acid (H_2CO_3). Carbon dioxide is present in air and dissolves by a series of reactions to form carbonic acid, the acid that makes carbonated beverages fizzy. The carbonic acid then reacts with water to make hydronium ions (H_3O^+) and bicarbonate ions (HCO_3^-). Here is the complete process as a chemical equation:

$$CO_2(g) + H_2O(l) \rightleftharpoons H_2CO_3(aq) + H_2O$$
$$\rightleftharpoons HCO_3^-(aq) + H_3O$$

Carbonic acid is a very weak acid, so carbonic acid produces very few hydrogen ions when dissolved in water. With the current levels of carbon dioxide in the atmosphere (about 370 ppm as of 2006), rain water is naturally acidic, with a pH of about 5.6 (this corresponds to 2.3×10^{-6} moles/liter of

H⁺ ions). This slight acidity is a natural part of the transport of carbon in the environment.

pH and Total Acidity

There are a number of ways to measure the pH of a solution. The most commonly used method in the laboratory is to use a device that produces an electrical potential (voltage) that is proportional to the concentration of hydrogen ions, which we express as pH. This device is first calibrated using solutions whose pH values are known very accurately. A less expensive way to determine pH is to use test papers, which contain dyes that change color depending on the pH. These are not as precise as a pH meter, but they are an inexpensive and portable way to get an approximate result. (You may be given the opportunity in one of the laboratory exercises included in this manual to explore these techniques for yourself.)

There is a second way to describe the acid or base present in a solution. We noted earlier that some acids do not liberate all their hydronium ions, and that the measurement of pH determines only the free ions. The total acid present in a solution can be measured by a process called *titration*. A titration is a chemical reaction in which the acid sample to be measured is reacted with a base solution of exactly known concentration. The base is added until the solution is neutral, at which point all the acid has been reacted. At this point, the moles of base added are equal to the moles of acid in the original sample being titrated. Unlike pH measurement, which measures only the hydronium ions that are free in solution, a titration measures all the hydrogen ions that will react with a base solution.

The differences between strong and weak acids are most notable when you compare the pH of solutions of strong and weak acids with the total concentration of acid found by titration. Because pH is a measure of the fraction of acid that has been converted to hydrogen ions, the pH of a weak acid will be higher (less acidic) than the pH of a strong acid of the same concentration. If you compare the ratio of free hydrogen ions to the total amount of acid, you will find that strong acids give a ratio very close to 1. Essentially all the acid present has donated its hydrogen ions to the solution. Weak acids give very small ratios (typically less than 0.10) because they do *not* donate many of their hydrogen ions to the solution.

Parts per Million

The term *parts per million* (ppm) is the most common unit in environmental chemistry. You may be familiar with the use of percent (%). A percentage can be calculated from the ratio of any two quantities having the same units and then multiplying by 100. You could also think of percent as part per hundred, or pph. A 1 percent sodium chloride solution means you have 1 part of sodium chloride in a 100 parts solution. We can think of 1 ppm as 1 part in 1 million parts. So a 5 ppm solution of sodium chloride could be interpreted as 5 grams in 1 million grams of solution. The density of water is 1.0 grams per milliliter, so a 5 ppm solution can also be thought of as 5 grams in 1 million milliliters of solution. If we divide both grams and milliliters by 1,000, this gives us milligrams per 1,000 milliliters, which is the same as milligrams per liter. We will use this as our definition of ppm: 1 ppm = 1 mg/L.

How Does Rain Become Acidic?

Acidity Due to Sulfur Pollution

Both oil and coal contain sulfur. In coal, the sulfur is primarily due to the presence of the mineral pyrite, or iron (II) sulfide, FeS. When the coal is burned in the presence of oxygen (O_2), pyrite reacts according to the following equation to form iron (III) oxide (Fe_2O_3) and sulfur dioxide (SO_2):

$$4FeS + 11O_2 \leftrightarrow 2Fe_2O_3 + 8SO_2$$

Other metal sulfides are also sources of SO_2. The most notable one is nickel sulfide, NiS, the principle ore used in the production of nickel for stainless steel. Nickel ores are "roasted" and then heated in air, which releases the sulfur as sulfur dioxide and converts the ore to an oxide.

$$2NiS + 2O_2 \rightarrow 2NiO + 2SO_2$$

Metal smelting is another significant cause of acid precipitation. In Sudbury, Ontario, the tallest smokestack in the world was built to disperse the pollution caused by the nickel industry. This smokestack was over 1,000 feet tall. It protected the local population from the sulfur dioxide pollution—and sent the pollution all the way to Scandinavia.

Sulfur is also present in all petroleum products. When these products are burned for fuel they also produce sulfur dioxide. The actual amount of sulfur in fuels varies considerably from much less than 1 percent to as much as 10 percent depending on the source, but sulfur is always present. The geological processes that convert biological material from dead plants and animals into fossil fuels—coal and oil—leave the sulfur from the organisms in the fuel.

The production of sulfur dioxide is something you may have experienced directly. Have you ever been stuck in traffic behind a car or bus that smelled somewhat like rotten eggs? That offensive smell is sulfur dioxide. Power plants all give off some sulfur dioxide. You may experience a burning sensation in your nose and throat when driving near a power plant if there are high concentrations of sulfur dioxide in the air.

Sulfur dioxide is quite reactive and toxic. Reactions in the atmosphere convert sulfur dioxide into sulfuric acid. The scheme shown here is typical of the process involved. Sulfur dioxide (SO_2) is first converted into sulfur trioxide (SO_3) by reaction with atomic oxygen (1), ozone (O_3) (2), or hydroxyl radicals ($OH^{\cdot}$) (3).

$$SO_2 + O \leftrightarrow SO_3 \quad (1)$$
$$SO_2 + O_3 \leftrightarrow SO_3 + O_2 \quad (2)$$
$$SO_2 + 2OH^{\cdot} \leftrightarrow SO_3 + H_2O \quad (3)$$

Sulfur trioxide reacts rapidly with water in aerosols to become sulfuric acid, H_2SO_4, a strong acid that has a dramatic effect on the pH of a solution:

$$SO_3 + H_2O \leftrightarrow H_2SO_4$$

Acidity Due to Nitrogen Pollution

The other source of acidic compounds in rain is also related to burning fuels. All combustion processes that use air as a source of oxygen produce pollutants, referred to as NO_x (read this as the three letters N O X). This is a generic term that includes a number of nitrogen oxides including NO, NO_2, and N_2O_4. If you look across any city on a clear day and see a yellow-brown haze in the air, NO_x pollution is the cause. These nitrogen oxides are produced through a complex series of reactions starting with nitrogen and oxygen gases. The reaction here shows the formation of an NO molecule with a free unpaired electron. This is referred to as the nitric oxide radical or $NO^{\cdot}$ radical,[6] and it is extremely reactive:

$$N_2 + O_2 + Heat \leftrightarrow 2NO^{\cdot}$$

This nitric oxide radical can react in a number of ways. It reacts with oxygen as follows:

$$2NO^{\cdot} + O_2 \leftrightarrow 2NO_2^{\cdot}$$

Due to the presence of an unpaired free electron, the highly reactive nitrogen dioxide radical $NO_2^{\cdot}$ will react with another nitrogen dioxide radical to form N_2O_4. However, because there is much more oxygen in the air than nitrogen dioxide, the radical is much more likely to react with oxygen. The reactions that occur depend in part on temperature and whether there is available sunlight.

---

6. The dot (˙) in the chemical formula is the way chemists indicate that a molecule has an unpaired electron. These chemicals are called radicals. They are highly reactive substances.

During the day, the nitrogen dioxide radical reacts in the presence of sunlight to release atomic oxygen, which then reacts with oxygen to form ozone (O_3).

$$NO_2\cdot + Light \leftrightarrow NO\cdot + O\cdot$$

$$O\cdot + O_2 \leftrightarrow O_3$$

The production of ozone in the atmosphere is more complex than is shown here, but this reaction serves to show how ozone can be formed in the presence of sunlight.

At night there is no sunlight to initiate the reaction, so a different series of reactions occurs. These reactions lead to the destruction of the ozone that has built up during the daylight hours and lead ultimately to the formation of nitric acid:

$$NO_2\cdot + O_3 \leftrightarrow NO_3\cdot + O_2$$

The combination of nitrogen trioxide radicals with nitrogen dioxide radicals leads to the formation of dinitrogen pentoxide, N_2O_5:

$$NO_3\cdot + NO_2\cdot \leftrightarrow N_2O_5$$

Dinitrogen pentoxide reacts with water vapor to form the nitric acid, HNO_3, a strong acid:

$$N_2O_5 + H_2O \leftrightarrow 2HNO_3$$

Another scheme that leads to nitric acid formation is a direct reaction of nitrogen dioxide with water in aerosol particles:

$$2NO_2\cdot + H_2O \leftrightarrow HNO_3 + HNO_2$$

This reaction also produces nitrous acid, HNO_2, a weak acid that is very quickly oxidized to nitric acid. Thus, all oxides of nitrogen emitted into the air ultimately become acid rain.

What Can Be Done about Acid Rain?

To solve the problem of acid precipitation, the amount of SO_2 and NO_x pollution must be decreased. Several strategies can be used to achieve this goal, but in dealing with environmental issues there are always problems associated with every solution.

Solution	Burn only low-sulfur coal or clean coal to remove sulfur before it is burned.
Problem	This type of coal is expensive and not available in Europe.
Solution	Install scrubbers on coal-fired power plants to remove SO_2 pollution.
Problem	The technology is expensive and produces waste products that must be disposed of.
Solution	Remove sulfur from petroleum fuels at the refinery.
Problem	The process is expensive and is impossible at some refineries without reconstruction.
Solution	Reduce NO_x emissions from automobiles by catalytic converters and the use of lean-burn engines.
Problem	The devices require the more expensive unleaded gasoline that is not available in Europe and also raise the cost of cars.
Solution	Reduce NO_x emissions from power plants using new burners, scrubbers, and fuel additives.
Problem	The technology has yet to be developed.
Solution	Use more natural gas to make electricity because it is more efficient and produces little pollution.
Problem	Natural gas is not available in Europe in great enough amounts.
Solution	Build more nuclear power plants, which produce no air pollution.

Problem   Nuclear plants are expensive to build, and the waste generated presents disposal problems.

Solution   Reduce the total energy used so less fuel is burned.

Problem   Reduction requires capital investment and shifts in the power sector.

Details of Sulfur Dioxide Reduction Approaches

All the measures we have listed will reduce the pollutants associated with acid rain, but none of them are cheap or politically easy. Low-sulfur coal, which is found primarily in the western United States, is not readily available on a global basis. Transportation of low-sulfur coal to places where it is needed has both environmental and economic costs that must be considered. Eliminating jobs from areas that are financially dependent on mining high-sulfur coal can be economically and politically devastating. England and West Germany both have extensive mining industries, and the miners' unions have considerable political power.

Eastern Europe is highly dependent on very low quality coal called lignite. This material is very high in sulfur (up to 15 percent) and low in total energy content. Thus, more lignite is needed to produce a kilowatt of electricity than would be needed for higher quality coal. The lignite burned in Eastern Europe could be replaced by coal from the West that contains much less sulfur, but the Eastern European economies lack hard currency, so they would need financial help to accomplish this. Also there are no good transportation corridors between the East and West as a result of the Cold War.

Some western European countries also use lignite as a major fuel. The challenge in transnational negotiations is to find economically feasible ways to replace these high-sulfur fuels. Considerable money has been spent on developing technologies to remove sulfur from coal before it is burned. Washing coal to remove the iron pyrite particles that contain the sulfur can be reasonably effective in reducing but not eliminating sulfur. The cost of washing coal can be as high as $1,000 for every ton of sulfur removed.

Flue gas desulfurization (scrubber) (FGD) technology is well developed, but scrubbers are expensive to install and operate. The initial cost of installing the scrubbers is relatively high, but the overall cost is only about $500 per ton of sulfur removed. This method removes more sulfur than the washing method. Research on nonpolluting technologies for burning coal is ongoing. Most scrubber systems work by passing the exhaust gases from the coal through wet limestone, $CaCO_3$. The reaction converts the limestone, $CaCO_3$, to calcium sulfite, $CaSO_3$:

$$SO_2 + CaCO_3 \rightarrow CaSO_3 + CO_2$$

The product is a wet slurry that is heavy and difficult to dispose of. This product can be further converted by reaction with oxygen to calcium sulfate, $CaSO_4$, commonly known as gypsum, which is a potentially useful building material. However, the product is contaminated by fly ash, a residue from the coal, and is thus not always suitable for reuse.[7]

A new technology called fluidized bed combustion places the limestone directly in the boiler in the form of a fine powder. This process is used in some new power plants and offers one of the best technologies for coal burning. In this approach, the limestone is converted into calcium oxide (CaO) by the heat of the boiler. This reacts directly with the SO_2 in the boiler to form calcium sulfite, $CaSO_3$.

---

7. Coal power plants emit more radiation each year than nuclear power plants due to the presence of radioactive elements in the coal that are either released in the fly ash or remain in the larger pieces of ash, which must be discarded. The ash from coal burning is, therefore, not something one wants to place close to people.

Existing boilers cannot be retrofitted to make use of this technology. While fluidized bed combustion still produces large amounts of gypsum waste, the waste produced is dry:

$$CaCO_3 + Heat \rightarrow CaO + CO_2$$

$$CaO + SO_2 \rightarrow CaSO_3$$

The sulfur in petroleum is also a problem that must be addressed. Crude oil that is low in sulfur is called "sweet" and sells for a premium price. Refineries prefer to work with low-sulfur fuels because high sulfur content can damage catalysts in the refinery. Catalysts are used to break the long hydrocarbon chains in crude oil into shorter molecules for gasoline. Because the amount of crude oil in the world is limited, if Europe decides to require low-sulfur fuel oils, they will need access to enough of this sweet crude, which will increase its price. Other countries will then buy the high-sulfur crude oil, resulting in little change in the overall global sulfur emissions. Technological development to improve removal of sulfur from fuels is an ongoing area of research.

The Formation of the EU and Transnational Negotiations

To students in the twenty-first century, the world of just a few decades past may seem hard to imagine. Many of you may have traveled in Europe and enjoyed the Euro being the common currency of most of the continent and the seamless travel from one country to the next without any border control or inspections. But in 1979 when this game begins, this was not yet the case. Nations did not come to accept the idea of the EU quickly or easily.

Environmental negotiations in the 1970s and 1980s were dominated by national interests and priorities. There was a vast disparity between the economies of the richest and poorest members of the EEC. These differences are a central issue in the acid rain negotiations. No nation was willing to see an economic recession at home in order to protect the environment of other countries. The EU, open borders, and a single currency were far in the future. However, if these negotiations failed, the dream of European unity would certainly never have become reality.

Our discussion will be an alphabet soup of organizations, and you need to keep them straight, so we will provide a brief history of these organizations and their activities.[8]

Organisation for Economic Co-operation and Development

The Organisation for Economic Co-operation and Development (OECD) is a Paris-based think tank focused on promoting economic development. Its study of environmental issues was based on the idea that environmental degradation was a limiting component of economic growth. In the late 1970s, the OECD published the results of their research as *The Long Range Transport of Air Pollutants*.[9] This document provided the first independent assessment of the net import and export of pollution from a large number of countries. The study was based on extensive measurements, but its limitation was the fact that the results were given an uncertainty of ±50 percent. This allowed the major polluters to claim that their contributions were vastly overestimated in the study. (England may make this argument in the game.) Errors could have come primarily

---

8. This information is drawn almost exclusively from G. S. Wetstone and A. Rosencranz, *Acid Rain in Europe and North America: National Responses to an International Problem* (Washington, D.C.: Environmental Law Institute, 1983).

9. Organisation for Economic Co-operation and Development, *The OECD Programme on Long Range Transport on Air Pollutants: Measurements and Findings* (Paris: Organisation for Economic Co-operation and Development, 1977); *The OECD Programme on Long Range Transport of Air Pollutants: Summary Report* (Paris: Organisation for Economic Co-operation and Development, 1977).

from shifting weather patterns when the specific measurements were made.

The OECD also published a major work on cost-benefit analysis of pollution control. They also published recommendations on pollution control strategies including the following:

1. Flue gas desulfurization (FGD) should be used on all power plants greater than 200 megawatts (MW).
2. Fuel desulfurization should be used on all oil at power plants without FGD.
3. Plants burning lignite (high-sulfur, low-quality coal) greater than 100 MW should use FGD.
4. All hard coal should be washed to reduce sulfur and particulate emissions.
5. Only low-sulfur fuel should be used where FGD is impractical.[10]

The OECD analysis suggested that countries should spend 1 to 2 percent of their GDP on pollution control and that the overall benefits would be at least equal to this cost and possibly as much as 3 to 5 percent of their GDP.

The OECD had no official position in international politics. Its studies were not in any way binding, and it had no power over the member countries other than the power of the logic and quality of its findings.

European Economic Community

The European Economic Community (EEC) is composed of the 10 nations of the European Common Market. The formation of the EEC was an effort to enhance the economic power of the member states as a counterforce against the economic power of the United States, Japan, and the Soviet Union. The long-term goals of the EEC included the possibility of a common currency and the removal of barriers to trade and travel within the community.

During the period of the game, the EEC has already developed a central governing body, the Commission, which seeks to establish policies for all member states. The four main countries of the EEC—*Britain*, *West Germany*, *France*, and *Italy*—exercise the majority of power in the EEC because they have twice the representation of the other six nations, *Belgium*, *Greece*, *Luxembourg*, *Ireland*, *Denmark*, and *the Netherlands*. The Commission proposes actions for the entire EEC, but they must be approved unanimously by the Council of Ministers before they are sent to the member nations for further approval. The requirement of unanimous approval means that each nation has veto power over any new proposal. The political mechanisms of the EEC would seem to make it difficult to pass any proposal that is even minimally controversial.

France has been a major force in the development of the EEC. The French are eager to see European power grow as a counterbalance to the United States and the Soviet Union. The desire for unity within the EEC and the drive to build the Common Market is a major factor pushing reluctant countries to accept pollution controls which may not be completely in their own self-interest. The assumption is that any sacrifice in the area of pollution control will be offset by the other advantages of the Common Market. (France will take the lead in pointing out this fact whenever other countries drag their feet.) If the Common Market does become a single economic entity, any nation that does not follow its rules can expect to pay a large price to be allowed to export its products to the countries in the Common Market.

Great Britain is by far the most reluctant nation in the EEC. The idea of the Common Market is not popular with either the government or the general population. The British value their currency and their economic independence. However, if they fail to participate in the EEC, and it is successful without them, then Britain will

---

10. Wetstone and Rosencranz, *Acid Rain in Europe*, 137–38.

become marginalized. They are too far from the United States and Canada for any type of common economic community with those countries. They would also lose easy access to continental markets. As a result of these conflicts, the other EEC members may have a difficult time influencing Britain with threats of failure of the EEC. A more successful approach will probably be to have them consider the potential benefits.

UN Economic Commission for Europe

The UN Economic Commission for Europe (UNECE) became the focus of discussions on transnational pollution for a variety of reasons. Using the United Nations provided access to a larger range of counties than the EEC, which included only 10 nations. Norway and Sweden, the main plaintiffs in the environmental issue, were not part of the EEC, nor were the nations of Eastern Europe who were the major sources of pollution. UNECE included not only all the western and Eastern European countries, but also included the North American countries who were exporting pollution to Europe. Thus, UNECE seemed the only choice for serious discussions of a transboundary treaty.

During the period of the game, the Scandinavian countries fear that an agreement through UNECE will lack any real teeth, but they see it as their only choice. The EEC nations favor UNECE for the very same reason—anything it decides on will lack teeth. The Soviet Union and its satellites have been party to various negotiations through UNECE in the past, with very little to show for it. But they hope that these negotiations will establish a precedent for their dealings with the West and place them on a more equal footing.

Council for Mutual Economic Assistance

The Council for Mutual Economic Assistance (CMEA) was an organization of Soviet Bloc nations conceived as a communist counterpart to the EEC. That is, the CMEA is an organization composed of countries still under the sway of the Soviet Union.

The members of the CMEA could hardly be considered to have full sovereignty, as demonstrated by Soviet military invasions when they attempted to exercise independence from the USSR.

Before Geneva, the CMEA had never been a party to international agreements with the West. It is also not clear that the member nations wish to give it the authority to negotiate or act on their behalf. The issue is one of national sovereignty within the organization—the individual members of the CMEA may resist granting the CMEA too much power over them, both as an assertion of national sovereignty and as a way to resist Soviet domination.

Nature of International Agreements

Treaties between sovereign states are by their very nature voluntary. Compliance cannot be compelled easily; both public and political pressure are the only tools to ensure compliance. For any nation to comply with an international agreement, there must be general agreement that the principles behind the treaty are sound and that the mechanisms of the treaty are fair and reasonable for resolving conflicts that may arise.

After a nation enters into an agreement on the international stage, the implementation of the treaty remains entirely under the control of that nation. It is possible for such agreements to include direct payments between countries to support the objectives of the treaty. Thus, a wealthy country could pay part of the costs for a poor country to comply with an agreement. It is also common for treaties to include technology transfer between countries.

International treaties can also establish international markets and define their operation. Thus, the World Trade Organization (WTO) is an international body that attempts to provide a level playing field for international trade. It provides a mechanism for reviewing conflicts and allows nations to retaliate with tariffs or other measures when the WTO finds that a nation is violating the agreed-upon principles.

International treaties are normally negotiated at meetings such as the ones that form the core of this game. However, treaties that are signed at these meetings do not take force until they are ratified by the individual governments. Most such treaties take effect when the number of ratifying countries reaches a specific level, usually between and 50 and 75 percent. Clearly, if only a few countries actually ratify a treaty, there is little value in complying with it. Ratification by a large majority increases political pressure on the remaining countries to also ratify the treaty. In some cases, further negotiations and modifications are required to get a few holdouts to ratify.

Tools of Public Policy

There are a number of approaches that governments can take to control something like air pollution. Students in this game may consider what the best approach would be for their own role, based on the type of political system and the type of pollution to be controlled. However, for purposes of the game, you should concentrate on the percentage reduction targets in your role sheet and not become distracted by the mechanisms to be used. These will almost certainly vary from one country to another. Trying to define a specific mechanism will not be productive in the debates.

Command and Control

In the command and control approach, the government mandates specific actions and requires all industries and individuals to comply or face civil or criminal penalties. The mandate can be for the use of a specific technology. For example, a government could require that all coal-fired boilers over a certain capacity be fitted with FGD by a specific date. Similarly, a mandate could be established that all fuel oil sold in the country have less than 50 ppm sulfur.

A second approach to command and control requires the use of the *best available technology*. This creates a moving target of regulations in which a specific class of polluters is required to adopt whatever is best; as the technology improves, the mandate changes.

A third approach is to mandate specific emissions levels rather than the technology to get there. This approach allows each individual plant to determine the most cost-effective way to reach the mandated limit. The specific form of the limit or cap on emissions can be in the form of annual totals, daily totals, maximum per unit of time, or peak maximum. Thus, a plant could be allowed to emit 10 tons of SO_2 per week or 300 tons per year, or no more than 100 kg/hour or to have emissions that never exceed 100 kg/hour. This approach to regulation requires more monitoring and reporting than the requirement for specific technology, but it is also more flexible.

Polluter Pays

A second approach to pollution control is simply to establish a cost per unit of pollution that must be paid by the polluter. This creates an economic incentive to reduce pollution and can be a source of government revenue to cover costs of pollution. A power plant might be assessed a charge of $10 per kg SO_2. This approach also requires monitoring to insure compliance. It has the advantage of allowing economic forces to act on pollution, but it also makes it possible for companies with large profits, or which can simply pass the charges along to their customers, to continue polluting. The economic forces to remove pollution are only felt strongly if there is a competing source for the same product or service which pollutes less and can thus sell at a lower cost.

Cap and Trade

Cap and trade, popularized by the Reagan administration in the United States, combines some of the features of the two preceding approaches. The cap part is a mandated limit on total emissions of a specific pollutant for a nation or even for the entire globe. The cap can be gradually reduced over time to provide additional reductions.

Once the total emission cap has been determined, the right to emit the pollutant is allocated to all polluters on some basis. This takes the form of emission credits, which allow a specific amount of pollution. This allocation process must be perceived as reasonably fair to be politically viable. Usually this is done based on some fraction of current emissions.

A market is then established in which emission credits can be bought and sold. Once each industry knows its maximum allowed emissions, a market is established to allow emission rights to be traded. Thus, a company that is allowed to emit 100 tons of SO_2 but installed a scrubber to remove 90 tons can sell the 90 tons of emission credits on the open market. A polluter who finds it would cost more to modify their equipment to reduce pollution than to buy the credits would buy these credits. A polluter who finds it less expensive to buy those 90 tons of emission rights than to install equipment to remove 90 tons of SO_2 can purchase the credits as a way to stay within their limits.

The value of the emission rights will fluctuate depending on supply and demand. If the caps are gradually reduced, the cost of emission rights will rise as they become scarcer. This will make the installation of pollution control equipment more cost effective than buying emission rights for more and more polluters.

Another aspect of a market in pollution credits is that individual industries or countries as a whole may invest in pollution reductions in another country or industry in return for those pollution credits. Thus, if company A finds it will cost more to reduce their own pollution than to reduce pollution in company B, company A can pay for the pollution reductions at company B in return for the pollution credits. This will allow Company A to save money by taking the most economically feasible route to meeting their requirements.

The ability to implement these various systems depends in part on the relationship of government to industry. In communist countries, where the government owns most heavy industry, there is no way to implement this approach internally. The same applies to Great Britain, where the government owns the electric power industry. However, the government can look at where it is least expensive and most productive to install pollution-reduction equipment. On the international level, such a system could work, with one country paying another to reduce pollution instead of doing it themselves.

The lack of markets has led the communist countries of Europe to rely almost entirely on mandates to reduce pollution. In some ways this makes things simpler—the government decides what to do and then does it. They do not need to resort to more complicated mechanisms to reach the desired result. However, the advantage of the more complicated market approaches is that they can enlist private capital to pay for improvements. In the communist countries, all money to install scrubbers, for example, must come from the government itself.

The tools for nations to control pollution are not directly transferrable to international structures. Treaties between sovereign states are by their very nature voluntary. Compliance cannot be compelled easily; both public and political pressure are the only tools to ensure compliance. Although trade sanctions and tariffs could also be used to ensure compliance, these have not generally figured in these negotiations. For any nation to comply with an international agreement there must be general agreement that the principles behind the treaty are sound, that the mechanisms of the treaty are fair, and that there are reasonable mechanisms for resolving the inevitable conflicts that will arise.

GLOSSARY AND GUIDE TO ABBREVIATIONS

BAPMoN   Background Air Pollution Monitoring Network

BMU   Ministry for the Environment (West Germany)

CEGB	Central Electricity Generating Board (U.K.)	NO_x	Generic term for nitrogen oxide compounds
CFs	Chlorofluorocarbons	OECD	Organisation for Economic Cooperation and Development
CFL	Compact fluorescent lightbulb		
CMEA	Council for Mutual Economic Assistance	pH	Hydrogen ion concentration expressed as the negative logarithm. The term pH is used to describe the acidity of an aqueous solution expressed in numbers between 1.0 and 7.0, where 1.0 indicates a very strong acid and 7.0 indicates a very weak acid.
CO	Carbon monoxide		
CO_2	Carbon dioxide		
CSCE	Conference for Security and Cooperation in Europe		
EEC	European Economic Community		
EMEP	Evaluation of Long Range Transport of Air Pollution in Europe	OPEC	Organization of the Petroleum Exporting Countries
EPA	Environmental Protection Agency (U.S.)	PCBs	polychlorinated biphenyls
EU	European Union	ppm	Parts per million. This term is used to measure the concentration of very dilute solutions. A 5 ppm solution would mean that 5 milligrams are dissolved in 1,000 milliliters or 1 liter of solution. Note that 1,000 milliliters are equivalent to 1 liter.
FGD	Flue gas desulfurization		
FRG	Federal Republic of Germany (West Germany)		
GDP	Gross domestic product		
GDR	Democratic Republic of Germany (East Germany)		
GJ	Gigajoule (10^9 joules)	PRON	Patriotic Committee for National Rebirth (Poland)
GM	Gamemaster		
H^+	Hydrogen ion—chemical species that is used to measure the level or degree of acidity of an aqueous solution. The term aqueous refers to solutions that are dissolved in water.	RAIN	Reversing Acidification in Norway
		RAINS	Regional Air Pollution INformation and Simulation
		SO_2	Sulfur dioxide. This pollutant is associated with burning of coal and coal-fired plants.
K	Kelvin (temperature)		
LRTRP	Long-range transport pollutants	TEL	Tetraethyl lead
MP	Member of Parliament (U.K.)	U.K.	United Kingdom
MW	Megawatt—term used to describe a quantity of electricity.	UN	United Nations
		UNECE	United Nations Economic Commission for Europe
NERC	Natural Environmental Research Council (U.K.)	VOCs	Volatile organic compounds
NGO	Nongovernmental organization	WTO	World Trade Organization

2

The Game

MAJOR ISSUES FOR DEBATE

There are a number of issues that will arise during the game. The evolution of issues is a result of the ten-year span of the three international conferences that constitute the game.

Geneva Conference

The major issues for this conference are philosophical.

- What are the responsibilities of nations for the air pollution they export to other countries?
- Can any nation prove it has been harmed by pollution from another?
- If damage has occurred, what are the most efficient ways to solve the problem?
- In general, what is the relationship of humanity to the environment in which we live?

Helsinki Conference

By the time of this conference, significant research has been done in many nations to determine the specific problems that need to be solved.

- What measures must be taken to reduce damage by acid precipitation?
- What are the responsibilities of polluters to those they pollute?
- What are the responsibilities of wealthy nations toward poor nations?

Sophia Conference

The details of this meeting will depend on the outcome of the two previous meetings. However, it is expected that attention will move from control of SO_2 to NO_x pollution and smog. This will require a debate on leaded gasoline and lead pollution as well.

- Should catalytic converters be required on some or all new cars?
- Are there alternatives to catalytic converters that will reduce smog?
- Should leaded gasoline be phased out in Europe?

FRAMING THE ARGUMENT

There are three scientific issues on which arguments can be based.

1. Does the sulfur dioxide actually harm rivers and lakes?
2. Does airborne sulfur dioxide harm forests and farms? There are other possible candidates, including ozone.
3. If both 1 and 2 above are true, will domestic reductions of sulfur dioxide pollution provide improvements in the environment of foreign countries? This is the most complex and country-specific issue.

Economic issues are more complicated. It is difficult to quantify the exact benefits of pollution reduction. There is considerable uncertainty about whether the benefits of pollution reduction will be enough to cover the costs. Also, the benefits may occur in a different country than the one that paid for the pollution control. This is the primary argument for countries like West Germany and Norway to help pay for pollution control in Eastern Europe.

The three main arguments for pollution control can be stated simply as follows:

1. Pollution controls will reduce health care costs and save more money than they cost.
2. Agricultural productivity will increase and will help pay for the pollution controls.
3. Pollution reductions will save money due to less damage to buildings, bridges, roads, and other infrastructure.

The weakness of these arguments is that the exact benefits cannot be known until after the pollution controls are paid for.

RULES AND PROCEDURES

The game sessions are divided between two different types of meetings. The primary meeting is of the conference sponsored by the United Nations Economic Commission for Europe (UNECE). These sessions will be chaired by the UN Representative. The UN Representative will open the sessions, guide the debate, recognize the speakers, and work both within the debate and behind the scenes to obtain an effective treaty to reduce acid precipitation.

All countries participate equally in these sessions, and each country has one vote. In the case of countries with multiple representatives such as West Germany and Great Britain, the leader of the country delegation will cast the vote, and this vote must represent the majority position of the delegates from that country.

The second type of meeting will normally occur during the last game session and will include only the European Economic Community (EEC) countries. These meetings are generally short and allow the EEC to formulate a common position on any treaty that is under serious consideration. These meetings may discuss and vote on a proposed treaty, or they may formulate a new treaty proposal to take back to the UNECE meeting. The EEC rules require that every country approve any treaty that is to be binding on the EEC, so every country has an effective veto of any actions taken by the individual members. This means that the EEC meeting must approve the proposed treaty before it can be voted on in the UNECE session. Britain, West Germany, France, and Italy each have two votes in the EEC meeting; all other EEC countries each have one vote. However, the veto means that even the smallest country can stop a treaty even though the majority of the voting members have approved it.

There are three phases of the overall game: UNECE meetings in Geneva 1979, Helsinki 1984, and Sophia 1986. The two types of sessions can occur in each phase of the game.

ROLE OF MONEY IN THE GAME

Each faction in the game begins with a specific amount of money that can be used to implement a treaty for emission control. The amounts vary with the gross domestic product (GDP) of the country and the willingness of the population to support environmental investments. This means that the wealthy countries—the United Kingdom, West Germany, France, and Italy—have most of the money, but they would also bear the greatest cost due to their size.

The Organization for Economic Co-operation and Development (OECD) has recommended a total commitment of 1.25 percent of GDP over five years, which corresponds to an annual cost of 0.25 percent of GDP toward pollution reductions. They believe this is reasonable and will not cause significant damage to local economies or jobs. Each country receives a table of data developed using the models developed by the OECD as a guide. These tables show the cost of implementing various levels of pollution reduction. The values are five-year totals.

For some countries with rapidly growing economies, it will cost a lot of money just to maintain their current levels of pollution. These amounts may be much larger than the OECD recommendations for national spending. These countries will need financial support from wealthier countries to reach significant reductions.

Table 3 shows a typical example of what you will find in your role sheet for the country you are representing. The first column shows the various reductions in national emissions that might form the basis of a treaty. The second column is the percentage of GDP that would be required of the specific country to reach this level. The third column shows the percentage reduction in emissions for spending this much money—but without a similar reduction from the rest of Europe. In this case, the actual reductions in local pollution are limited due to the import of pollution from the rest of Europe. The last column shows the cost in U.S. dollars.

For the country featured in Table 3, a large expenditure is required to even hold pollution at the present level. And most of the pollution is imported, so even a 50 percent reduction in local emissions will only produce a 15 percent reduction—again, due to imported pollution. On the other hand, some countries such as the United Kingdom will reap most of the environmental rewards of any money they spend because little of its pollution is imported. Therefore, countries with primarily imported pollution have a large stake in getting everyone to agree to reduce pollution.

TABLE 3 Example of costs of emission reductions provided to each country

Percentage Reduction SO_2 Local Emissions within a Country	Cost as Percentage of GDP for the Country	Percentage of Reduction of SO_2 within a Country if the Same Money Is Spent but No Other Country Reduces Their Emissions: Difference Due to Imported Pollution	Cost in $ Million, 1985
0	0.58	0	$250
10	0.84	3	$400
20	1.10	6	$500
30	1.36	9	$650
40	1.62	12	$750
50	1.90	15	$850

WINNING THE GAME

Each national delegation has specific victory objectives defined on their role sheets. These produce specific numbers of Victory Points. In addition, there is the possibility of one or two Victory Points related to the money spent and the die roll. After the game is finished, the Gamemaster (GM) will tally the Victory Points for each faction and announce the winners at the Postmortem.

After the final game session, the die roll will be held. The value of the die and the amount of monies invested in pollution reduction by all countries will determine who receives Victory Points based on money spent. The rules determining who receives a Victory Point based on money spent are described in Table 4, the Victory Point scenarios.

The exact amount of money that constitutes a large or small investment and a high or low dice roll is not clear to you and is known only to the Gamemaster. This reflects the uncertainty of the costs and benefits of pollution reduction. A high dice roll signifies that the benefits of investing in pollution reduction outweigh the costs. In the case of a high dice role, countries that invested money at the Geneva meeting are awarded 2 Victory Points and those investing at Helsinki are awarded one Victory Point. This reflects the greater benefit of early investment and greater reduction in pollution in those countries.

OUTLINE OF THE GAME

Setup Sessions

The number of sessions will vary. Possible sessions are listed here.

> *Class 1. Introduction to Acid Rain.* Read the game book and "Geneva Conference News" in Chapter 5.
> Acids, Bases, and Acid Rain—Discussion of acid/base chemistry and acid rain.
> *Class 2. Lab Activity—Acids and Bases.* Class activity or full laboratory (optional).
> *Class 3. Environmental Philosophy.* Read the game book, Appendix 1, and Leopold, *Land Ethic*, pages 6-19, 25-36, 40-43, 70-82, 116-119, 188-202, and 237-264; Devall and Sessions, *Deep Ecology*, chapters 3 to 5.
> Study Questions in Appendix 10.
> *Class 4. Ecology.* Read Lovelock's *Gaia*—Discussion Questions.
> Study Questions in Appendix 10.
> *Class 5. Using Numbers to Make Arguments.* Environmental economics and cost-benefit

TABLE 4 Victory Point scenarios for money spent on treaty

Possible Outcomes	High Dice Roll	Low Dice Roll
Virtually all countries invested their money in pollution control.	Everyone wins, whether they invest or not, because pollution decreases across Europe.	Only countries that kept their money win; the benefits of investment by the countries that did invest are small.
Only a few countries invested their money in pollution control.	Only countries that invested their money win because they reap the benefits of their investment.	No one wins because total investment is not enough to reduce pollution.

analysis spreadsheet exercises, faction meetings.

Read Appendices 2 and 3.

Game Sessions

Each game session consists of plenary sessions of the UNECE. These sessions will be led by the UN Representative, who has been given the task of ensuring that a treaty is produced. The UN Representative is responsible only to the UN Secretary General (the GM in this case) and does not represent a specific country. The GM may replace a UN Representative who is not successful in leading the conference to a successful outcome. Conference sessions may be adjourned at any time for meetings of the various national factions. It will also be necessary to have the EEC representatives hold a joint faction meeting because they must all agree on any action before any member can vote for it. These EEC meetings will normally occur near the end of the second session of each conference.

The GM may also conduct a dice roll during each game session. These dice rolls are used to determine various political and environmental events that may or may not occur during the game.

Game Sessions 1 and 2: 1979 Geneva

The purpose of this first meeting is to formulate the Convention on Long Range Transport Air pollution. These sessions will address philosophical issues of environment and international cooperation. Papers will be written by each student presenting the philosophical and economic positions of their nation. Language of a general protocol must be approved with a framework developed for formulating specific emission reductions. Each faction should consider submitting a draft proposal as part of their initial writing assignment. Each nation has one vote in the UN Conference.

France holds the rotating presidency of the EEC during this period, and the deputy foreign minister of France will preside at any meetings of the EEC countries during these sessions. England, France, Italy, and West Germany cast two votes, and all other EEC countries cast one vote in EEC meetings. England, France, and West Germany also have veto power in these sessions.

At the end of Session 2, the EEC will announce its joint actions, and each individual country will announce what funds they will provide toward pollution reduction efforts (both internal spending and grants to other countries). The funds will be transferred to the GM.

Optional Laboratory exercises may be included between these sessions.

Game Sessions 3 and 4: 1985 Helsinki

Read "Helsinki Conference News" in Chapter 5.

This meeting will offer an opportunity to strengthen any measures approved in Geneva. Again, the UN Representative will preside over the plenary sessions. The results of any spending in 1979 will be announced by the GM. New data from scientific research and recent elections may also be available, which could change various national positions. Each faction should be prepared to propose specific changes to the agreement from Geneva or to propose a new agreement framework if the Geneva conference did not produce any agreement.

West Germany holds the presidency of the EEC, and the German foreign minister will preside at meetings of the EEC during these sessions.

At the end of Session 4, the EEC will announce its joint actions, and each individual country will announce what funds they will provide toward pollution reduction efforts (both internal spending and grants to other countries). The funds will be transferred to the GM.

Game Sessions 5 and 6: 1988 Sophia

The purpose of this meeting is to develop a protocol on nitrogen pollution. The results of previous efforts will be announced by the GM. Ongoing research results will be made available. Although the primary purpose is to agree on a nitrogen protocol, it may also be necessary to revise the sulfur pollution protocols, depending on the outcome of the previous sessions. Again, all factions should arrive at the conference with specific proposals to present.

The EEC presidency is held by Ireland for this session. England, France, Italy, and West Germany still have two votes each, but no country has a veto power. All countries are now bound by the majority decision.

At the end of Session 6, the EEC and the UN Conference will announce their actions, and each individual country will announce any remaining funds they will provide toward pollution reduction efforts (both internal spending and grants to other countries). The funds will be transferred to the GM.

Debriefing: Postmortem

The outcome of the game will be announced by the GM, and any secrets will be revealed.

What actually happened, and how well did it work?

What is the significance of these negotiations for other issues such as global climate change?

WRITING ASSIGNMENTS FOR THE GAME

Each student will write either two or three papers for this game. Every student should write a paper for the Geneva sessions. The nature of these papers will depend on your roles. Within factions, it is important to divide the research and writing so that each student focuses on a limited number of issues. Thus, one student from Great Britain might write only on the philosophical aspects of the argument, another on the scientific issues, and a third on the economic issues. This will allow for more depth in the writing and avoid superficial papers that all make the same arguments. Countries that are not part of factions generally have only one or two issues to address in their papers relevant to their own situation.

The students in the major factions may choose to divide the work for the second and third portions of the game so that some students concentrate on Helsinki and others on Sophia. This again will allow for more depth of treatment in the papers and avoid having several papers making the same points. These students would then write only two papers for the entire game, one for Geneva and one for either Helsinki or Sophia. But each faction should make sure that at least one paper is written for each phase of the game. Small countries will generally have specific issues for each game and will normally write a paper for each phase of the game dealing with their issues.

COUNTERFACTUAL ASPECTS OF THE GAME

The names of the characters in this game are all real people who held the positions specified. You are welcome to use the Internet to find out more personal details about any character, though in most cases this will not help you in the game. It is not clear that any of the named characters actually participated at the UN conferences in the game. (The actual list of participants has not been found by the authors.) However, the characters are people who would have been in a position to participate.

The ten-year span of the game encompasses many political changes in Europe. The people holding office in 1987 were often different from those who held office in 1979 when the game begins. Several of the characters changed positions in their governments during that time span, but the game excludes these changes from their roles because they are not relevant to the play of the game.

3 Roles and Factions

PUBLIC KNOWLEDGE ABOUT THE PARTICIPATING COUNTRIES IN 1979

The factions described below are relevant to the start of the game in 1979. You should not assume that these factions will persist throughout the game. All "Reacting to the Past" games include the possibility of changes in roles with time and in response to external events.

FACTION 1: MAJOR INDUSTRIALIZED NATIONS OPPOSED TO A STRONG TREATY

The group opposed to a strong treaty includes the major polluters of western Europe. Their large industrial economies have made them the wealthiest nations. They have been accused of exporting much of their air pollution to other countries while suffering little damage at home.

Great Britain (United Kingdom). The British delegation is led by the undersecretary of state of the environment. He will cast Great Britain's vote at the Conference. He is accompanied at the meeting by the chief alkali inspector, the head of a century-old agency responsible for regulating air pollution in England. The Alkali Inspectorate is known to have close ties to the industries it regulates and has been reluctant to publish any data on actual air pollution levels. However, the Alkali Inspectorate has a 100-year history of working to reduce air pollution. The third member of the delegation is the head of the Central Electricity Generating Board (CEGB). The CEGB is responsible for most of the electric power generation in the United Kingdom and is a government-owned utility. Public statements from the government indicate this group will not be friendly toward pollution reduction or environmental measures in general. There is possibly a fourth representative who is a member of the foreign service.

West Germany (Federal Republic of Germany, FRG). The federal foreign minister leads this delegation and casts West Germany's vote. The fact that the delegation is led by the foreign minister himself is seen as a clear indication of the importance the FRG places on these negotiations. He is accompanied in the delegation by both the director and assistant directors of the West German Ministry for the Environment (BMU). The BMU funds research and oversees protection of West Germany's valuable forest resources. Although the BMU delegates are both political appointees, they are also both scientists and will bring their knowledge of the pollution issue to the discussions within the West German delegation as it seeks to define its position on the issues.

FACTION 2: COUNTRIES SEEKING A STRONG TREATY

The countries who want a strong treaty claim to suffer primarily from air pollution imported from the major industrial nations. They want to reduce this pollution based on the UN Statement for 1972, which makes all nations responsible for the damage they cause with their exported pollution.

Norway. This delegation consists of the two cabinet ministers most involved in pollution issues, the minister of the environment and the minister of industry. They will make the case for pollution control most strongly because the damage attributed to acid rain appears to be most evident in Norway. They certainly share a common position that pollution reductions are needed, but they may not always agree with each other on the specifics due to their different portfolios in the government. They will, no doubt, be the first to speak at the UN Conference to voice their grievance against the major industrialized countries, whom they accuse of producing the pollution that they experience in Norway. They can also be expected to produce the strongest draft treaty on air pollution.

Sweden. The Swedish delegation is led by the director of the ministry of agriculture, the agency that sets environmental policy in Sweden. He is accompanied by the director of the Swedish Society for the Conservation of Nature, who is possibly the leading environmentalist in Sweden. The Swedish delegates will join the Norwegian delegates in calling for strong and quick action on this issue.

France. This delegation includes the deputy foreign minister and the ministers for the environment and for energy. France is widely recognized as the prime mover in the European Common Market and will press for a treaty on air pollution, even though France has not suffered significant damage itself. The motivation for this is not altruistic. The French have made a large commitment of national resources to nuclear power. This was done to protect their economy from a repeat of the 1973 oil embargo by the Organization of the Petroleum Exporting Countries (OPEC) nations, which decimated France's economy. Having invested in clean but expensive nuclear power plants now under construction, France feels it will be at an economic disadvantage in the European marketplace if its major competitors, West Germany and Great Britain, are free to use less expensive coal to power their industries. It is clear that the primary driving force in France's position at the conference is to provide a level economic playing field rather than a real commitment to reducing pollution in Scandinavia. However, they can be counted on to work closely with the Scandinavian faction. The French are also a main driving force for forming and strengthening the EEC. They see the ability to get an agreement on air pollution as an important first step to their more ambitious goals of a united European economy.

United States. The United States has already passed laws to reduce air pollution and will generally provide information on the successes and limitations of these laws.

FACTION 3: EASTERN EUROPEAN NATIONS

The Eastern European faction consists of countries on both sides of the Iron Curtain: Austria and Finland from the West, and the communist governments of Czechoslovakia, Poland, and Hungary from the East. Both Finland and Austria have close economic ties on both sides of the Iron Curtain, and they will argue that no treaty will be effective without the inclusion of the Soviet Bloc nations. All members of this group are eager to reduce their air pollution, but they lack the financial resources to do so. Their participation in any treaty will be contingent on getting the faction of industrialized countries to provide financial support.

Austria. The foreign minister of Austria has made it clear in his public statements before the conference that Austria, which receives much of its pollution from the East, needs to have a treaty that includes all of Europe, not just the Western democracies. Signing a treaty with only the West will do them little good. The Austrian economy is the strongest one in this group, and they will probably contribute financially to the effort if the industrialized countries can be convinced to pay the majority of the costs.

Finland. The Finns have a weak market economy and are very much in the shadow of the Soviet Union. The Finish government has agreed to participate in this meeting but has made it clear in their public statements that they are very concerned about the impact of expensive pollution controls on their economy. The OECD analysis makes it clear that it is not in the economic interest of Finland to spend any money on pollution controls. Therefore, they will align themselves with the communist countries in refusing to sign a treaty without financial support.

Czechoslovakia, Poland, Hungary, East Germany, USSR. The communist countries have almost no freedom of independent action from the Soviet Union. They are all members of the Council for Mutual Economic Assistance (CMEA), but the individual countries may resist having the CMEA viewed as a treaty signatory. This would be seen as reducing the limited sovereignty of the individual nations. This fact may complicate negotiations at some point. For now, each nation is known to suffer from some of the worst air pollution in all of Europe. There is some internal political pressure to address pollution because of the massive damage to health and the environment. What is lacking is the hard currency to purchase the needed technology. None of these countries can work independently of the others or the Soviet Union, which may act behind the scenes to limit their actions. Note that East Germany (German Democratic Republic, GDR) is not represented in this game due to our inability to locate the relevant economic data.

FACTION 4: LESS DEVELOPED NATIONS

The less developed countries are not major polluters, and their physical locations on the south and west side of Europe spare them from receiving pollution from other nations. Thus, they have little incentive to participate in a treaty and little or no resources to spend on a treaty that will require major investment on their part. They also resent the idea that the major industrial powers, who have gotten rich by polluting the environment, now want to deny that same avenue of development to the poor countries. There is a basic issue of fairness in their argument.

Spain, Ireland, Greece, Italy. Each of these countries is represented by their foreign minister.

4

Core Texts and Supplemental Readings

GENEVA CONFERENCE NEWS

Articles from *Nature* are summarized to provide context and information for the game.

E. Lawrence, "OECD Urges Environment Policies," *Nature* 252 (1974): 263.

A meeting of the Organisation for Economic Co-operation and Development (OECD) was held in Paris and attended by ministers from 24 countries.

Recommendations of the Environment Group of OECD put forth several concepts to formulate rational and coherent environment policies throughout the OECD member nations.

The concepts proposed are summarized below:

- *Polluter Pays Principle.* Polluting industries will not be given state aid to control their pollution issues but will be held accountable to meet accepted standards. The thinking is that this will make the cost of goods produced by polluting industries more expensive because they will bear the cost of their pollution. Consumers will prefer the less expensive products.
- *Code of Conduct.* Transfrontier pollution is pollution that is exported to other countries. The code of conduct proposes that transboundary pollution must not exceed that permitted within the polluting country. For example, Britain's coal-fired plants give off sulfur dioxide that is carried by wind to Scandinavia where it causes acid rain. Acid rain contributes to damage to forests, marble, and fish. This principle raises various complicated legal issues about liability to prosecution. Interestingly, Europe follows Roman law, but Britain follows common law, which works on a precedent system rather than specific legislation.

H. Leivestad and I. P. Muniz, "Fish Kill at Low pH in a Norwegian River," *Nature* 259 (1976): 391–92.

This article summarizes the decline in freshwater fish populations in southern Norway associated with increased acid levels in rivers and lakes. Specifically, the decreased populations of salmon and trout are thought to be due to the emissions, oxidation, and long-distance transport of air pollutants, especially sulfur dioxide. The same problems have been observed in the United States, Canada, and Sweden. In addition, the physiological effects of acid rain on fish are described.

Massive fish kills due to low pH (high acidity) have not been well documented, but the massive fish kill in the Tovdal River in southern Norway in the spring of 1975 was studied in detail. The physiological changes due to high acid concentration were documented in salmon and brown trout. This river was a major source of salmon and brown trout; however, in the spring of 1975 the river no longer had any salmon but still had brown trout. At the time of the fish kill, ice still covered parts of the river. Thousands of dead trout covered the bottom of the river. It should be noted that there is no industry or high-density population in the river valley. The fish kill was thought to be associated with the release of pollutants from the start of the annual spring snowmelt.

An investigation of the blood chemistry showed that fish associated with areas with high concentrations of dead fish had plasma chloride levels that were much lower than fish tested from areas with low numbers of dead fish. Low plasma chloride levels are associated with fish "stressed by high acidity levels." A low level of chloride ions is accompanied by low levels of sodium ions. When the "stressed" fish were transferred to tanks with lower acidity (pH about 6.0), after two weeks the previously "stressed" fish returned to healthy levels of sodium chloride. This indicates that high acidity is the cause of damage to fish in the Tovdal River. It was concluded that fish kept in increased-acid waters lose their ability to regulate plasma sodium and chloride levels. Imbalance of these ions damages both the conduction of nerve impulses and metabolism.

In conclusion, the acidity (as measured by pH) below which fish populations are affected depends on some additional factors. Salmon seem to be more sensitive than trout. Newly hatched fish are more sensitive than adult fish.

The total concentration of dissolved ions in rivers and lakes changes the sensitivity to low pH. It has been observed that waters with fewer dissolved ions (lower electrical conductivity measurements) have lower numbers of fish. Waters with lower acidity (higher pH) but more dissolved ions also show lower fish populations. The key observation was that the first phase of the spring snowmelt produces a pH shock to rivers and top layers of lakes. This reduces both the total ion content and the pH, both of which increase the acid stress on fish.

M. Duckenfield, "Controlling the Pollution Trade," *Nature* 264 (1976): 107.

Norway has passed legislation requiring all industries to use low-sulfur fuel, and Sweden begins a program next year to reduce sulfur emissions to 1950 levels. These moves are designed to allow the Scandinavians to increase pressure on other polluters by dealing with their own pollution.

The new regulations in Sweden result from extensive studies sponsored by the Ministry of Agriculture. The study found that 63 million tons of sulfuric acid compounds were emitted in 1973 in Europe. This is a major increase from the 25 million tons generated in the years from 1910 to 1950. Of these emissions, 38 million tons of sulfur were attributed to sources in Eastern Europe, and the rest came from western Europe. It was predicted that by 1985, 74 million tons would be generated, with 27 million tons coming from the West. It should be noted that Sweden burned about 19 million tons of fuel oil and emitted about 800,000 tons of sulfur. The growing economy is expected to emit 1 million tons by 1985 if no action is taken. The new Swedish regulations will cut emissions to 300,000 tons.

The new regulations beginning next year will limit sulfur in light fuel to 0.5 percent sulfur, and this value drops to 0.3 percent in 1981. The limit for heavy oil is set at 1 percent sulfur in part of the country south of Stockholm. A limit of 2.5 percent sulfur for all fossil fuels, including coal, will also be established. This plan will cost about $130 million.

Unfortunately, this plan will only keep the acid rain problem stable in the worst parts of the country. Because less than 25 percent of the pollution in these areas comes from Swedish sources, the plan cannot really reduce the problem to a significant extent without cooperation from the rest of Europe. Sweden is pressing for a 90 percent reduction in total emission from western Europe in the next 10 years.

Norway

The ability of Norway to reduce its acid pollution is much more limited than in Sweden because less than 10 percent of the pollution is generated in Norway. Most of the pollution is transnational. Even in the area near Oslo, which contains most of the heavy industry in Norway, the local contribution to the problem is only 30 to 40 percent. The thin, granite-based soil in much of the country cannot neutralize the acid rain, and the acid released into rivers and lakes by snowmelt during the spring thaw leads to the huge fish kills of hatching and spawning trout and salmon. Many of the lakes and rivers in the southwest of Norway no longer have any fish.

The Norwegian government is funding 60 scientists at a cost of close of $12 million to investigate the effects of acid rain. Research shows that not only are fish dying in great numbers, but also acid rain is having a negative effect on forest production due to the depletion of essential nutrients from the soil. In urban areas, acid rain is causing the corrosion of metals and stone.

To draw international attention to the need for control of acid rain, a conference organized jointly by the Norwegian Ministry for the Environment and the precipitation project (SNSF) was held in Norway and attended by representatives of 20 nations of the Common Market (all but Luxembourg), the Soviet Union, and three Eastern European nations.

The conference only produced an agreement that countries consider steps to reduce their emissions. However, a report by the Organisation of Economic Co-operation and Development (OECD) based on research into long-range transport of air pollution underscored how damaging acid rain is to Norway and Sweden. The report noted one twelve-month period in January 1974 when deposition rates as high as 1 ton of sulfate per km^2 were observed in southern Norway. This report should increase pressure for international cooperation on the problem.

W. Gries, "West Germany Energy Plan Published," *Nature* 267 (1977): 98–99.

The West German federal government has published an energy research plan for 1977–1980. The cost of this plan is $14.4 billion annually, and the major points are summarized here:

Nuclear energy—75 percent of the cost of the plan will be used for research and development in the fields of fast breeder reactors, uranium enrichment, and reprocessing and disposal of radioactive waste. Reactor safety is a key component of this work.

Rational energy use—About $230 million will be used to fund research into technologies that will use energy in a rational manner. These technologies include centralized heat supplies, reverse-cycle heating systems, and waste-heat recovery from industry and power stations.

Coal technologies—Improvements to coal-fired power stations to reduce pollution using new technologies is a key point in this governmental plan, and $325 million will be used to investigate coal liquefaction, improved mining techniques, and any other promising technologies.

New energy sources—Nuclear fusion and solar energy for heating and hot water will be investigated, and $300 million is expected to be available for research into these energy sources.

This new energy program was determined by the federal government by a cabinet resolution. The country's sources of energy:

Petroleum	52%
Hard coal	19%
Soft coal	10%
Natural gas	14%
Nuclear energy	2%
Others	3%

A breakdown of energy use by sectors in 1975:

Industry	36%
Transport	20%
Domestic	44%

The government believes the largest energy savings can be found in the domestic sector.

W. Barnaby, "Sweden," *Nature* 267 (1977): 99.
Recent report by Sweden's National Environmental Protection Board indicates that cleaning up polluted lakes is a very slow process. The lake pollution in the southern third of the country results from a combination of acid rain, sewage, and industrial and agricultural runoff.

The reduction of acid rain requires that fuels be burned without releasing sulfur compounds into the air. Spreading lime (an inorganic calcium compound, usually calcium oxide) onto the lakes has been tried as an interim solution to reduce the acid to safe levels. This measure is costly, and the toxic metals found in most commercial lime will poison the lakes if the required amount of lime is applied.

Another approach considered is reducing the runoff of waste products and fertilizer into the polluted lakes and streams. This only improves water conditions slightly.

There are multiple pollutants, including polychlorinated biphenyls (PCBs) and mercury, and the changes in ecosystems cannot always be attributed to specific causes. As a result of the uncertainties, it is hard to devise plans to remediate all the observed pollution problems.

K. Mallanby, "Million-Dollar Problem—Billion Dollar Solution?," *Nature* 268 (1977): 89.
Editorial—For the past 30 years, the solution to the problem of sulfur emissions has been solved by building taller and taller smokestacks. While these solve the local pollution problem, they simply move the problem elsewhere. The material spreads and travels until it eventually falls as dry deposition or is washed out as acid rain.

The recent Organisation for Economic Co-operation and Development (OECD) report attempts to estimate the amount of pollution imported and exported by each country. These estimates have very large uncertainties, up to 50 percent or more, and cover only a single year. Politicians who must make decisions about pollution control will probably misinterpret the data because they don't understand the uncertainty. Austria, Switzerland, Finland, Norway, and Sweden are notable because they import substantially more pollution than they export. On the other hand, the United Kingdom and Denmark are net exporters of sulfur pollution.

The United Kingdom is considered the largest emitter of sulfur in Europe, with emissions of roughly 1.8 million tons per year. More than half of these emissions are exported to the rest of Europe. In Norway, British sulfur pollution represents a larger fraction of the total deposited sulfur than Norway's own emissions. The effects of this pollution have devastated fish in lakes and rivers in southern Norway.

TABLE 5 Cost of nuclear and alternative energy sources

Energy Source	Capital Cost	Annual Operation	20-Year Total
Nuclear	$175	$3.60	$246
Storage systems	$90	$2.10	$132
Solar Panels	$85	$0.50	$95
Wind	$49	$0.80	$65
Wave	$125	$3.00	$185

Source: Ryle, 1977.

While the pollution levels in Norway are not any greater than in other parts of Europe, the effects have been greater due to the nature of the underlying soil and the fact that the pollution builds up in snow during the winter and is then released in a highly concentrated form in spring. While all of Europe suffers from damage to forests, agriculture, and buildings due to acid emissions, the effects on fish are most serious in Norway as a result.

The Norwegians will use the OECD report as another opportunity to demand better removal of sulfates. The United Kingdom will say that there is not enough evidence to say for sure that this is their fault. The question at hand is whether the United Kingdom should assume any responsibility for the effects of acid rainfall in Norway. The damage in Norway is on the order of tens of millions of dollars. Unfortunately the cost of a Europe-wide solution will be on the order of 1 to 10 billion dollars. Politically this is a very problematic situation.

A number of technological fixes have been proposed. One such proposal is that Britain provide Norway with limestone (calcium carbonate) to be added to the lakes during the time that the spring melt adds a high amount of acid to the water. This is referred to as the "acid flush" and is thought to account for the large number of fish kills in the spring. It is not clear whether this will work and what other problems might result. Although this is not the most appealing remedy to this problem, the Norwegian government should not simply reject it without serious consideration. Due to the high investment costs of removing pollution at its source and the generation of international ill-will, all solutions need to be considered.

M. Ryle, "Economics of Alternate Energy Sources," *Nature* 267 (1977): 111–17.
This is an extensive discussion of alternative sources of energy available in the United Kingdom, including wind, solar, and ocean wave. Data are presented for the relative cost of various energy sources in current use. These costs are useful in order to understand the current energy mix in the United Kingdom.

Table 5 compares the capital and operating costs of various fuel systems in the United Kingdom per gigajoule (GJ, 10^9 joules) of energy per year.

Storage systems are needed for all alternate energy systems to deal with daily and weekly variations in weather. They must provide 150 hours of storage to allow the supply to adjust to changing demand. To get the true cost of using only alternative energy, one needs to add the storage system cost to the cost of constructing and operating the alternative energy system. Thus, wind energy, in the absence of coal or nuclear backup, would be $132 + $65 = $197.

Annual operating cost of current power stations in the United Kingdom for 1974 to 1975 per GJ of

TABLE 6 Annual operating cost for 1 gigajoule of energy in the United Kingdom, 1974

Coal	Oil	Nuclear	Hydroelectric
$3.92	$6.6	$1.19	$0.30

energy per year are shown in Table 6. The values are in U.S. dollars per GJ of energy per year. Note that these values do not include the costs of plant construction. The cost of building a nuclear power plant is high and uncertain. The costs for nuclear also do not include the costs of storage and disposal of nuclear fuel. These are uncertain because they have yet to be done. Also nuclear plants must be removed at the end of their life span, the radioactive components must be disposed of, and the site must be decontaminated. None of these costs are known at the present time.

The author of this article argues that alternative sources can meet the energy needs of the United Kingdom in a nonpolluting way so long as a major investment is also made in energy storage technology.

GENEVA CONFERENCE: SCIENTIFIC REPORTS

The following technical papers are summarized for use in the game.

P. F. Chester, "Acid Rain, SO_2 Emissions, and Fisheries," in *Acid Precipitation: Presented at the 1981 Winter Meeting, Atlanta, Georgia, February 3, 1981*, Special Publication of the IEEE Power Engineering Society, no. 20, 13–18 (Piscataway, N.J.: IEEE Single Publication Sales, 1981).

Lakes in the Sorlandet region of southern Norway were studied to understand the reasons for many lakes in the region becoming fishless. The fishless lakes are more common at higher altitudes. The lakes above 200 meters were chosen for detailed chemical analysis. In addition to pH and

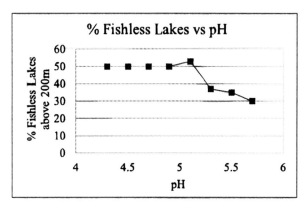

FIGURE 3 Effect of pH on percentage of fishless lakes in Norway. (Data from Chester, 1981, p. 14, figure 5.)

sulfate levels, the total ion concentrations and the calcium and magnesium from weathering rock were measured.

Hydrogen ions varied by a factor of 1,000 in these lakes, but sulfate varied by only a factor of about 2. The concentration of the cations calcium and magnesium showed the extent of rock weathering that had neutralized the acid inputs. If the primary acid input was from atmospheric deposition, the sulfate ions would correlate with the positively charged metal ions. However, this is not the case. The data provide evidence that other acids, probably carbonic acid from biological activity, is a major factor in acidification.

Figure 3 shows the relationship of fishless lakes to pH. The fact that the fraction of lakes that are fishless does not change below pH 5.0 and that there are more fishless lakes at pH 5.1 than at 4.3 shows that pH is not the only factor in whether the lakes have fish.

Figure 4 shows similar data for the relationship of sulfate concentration and fishless lakes. If the death of fish were due to sulfuric acid pollution in rain, one would expect the percentage of fishless lakes to increase as the sulfate concentration does. This is clearly not the case. From this, one must conclude that the input of sulfuric acid through acid precipitation is not the primary cause of fish death in these lakes.

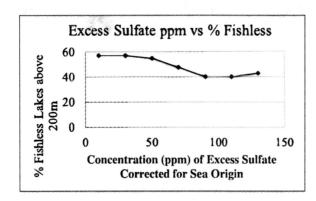

FIGURE 4 Sulfate concentration and percentage of fishless lakes in Norway. (Data from Chester, 1981.)

The data show an increase in the percentage of fishless lakes as the altitude increases. Higher altitude lakes are also characterized by lower calcium concentrations. Calcium is an important nutrient for all life. High altitude lakes with pH greater than 4.7 but low in calcium and sulfate had a higher percentage of fishless lakes than low altitude lakes with pH less than 4.7 but higher calcium and sulfate.

Data also are presented for the deposition and concentration of sulfate at two groups of Scandinavian weather stations over the period of 1955 to 1975. After spiking in 1963, both the deposition and concentration values declined and remain around the twenty-year average.

In conclusion, the relationship between fisheries and sulfur emissions is complex, and the present data do not allow any projection of the impact of reducing emissions on fishery health.

J. P. Nilssen, "Acidification of a Small Watershed in Southern Norway and Some Characteristics of Acidic Aquatic Environments," *Internationale Revue der gesamten Hydrobiologie und Hydrographie* 65, no. 2 (1980): 177–207.

Chemical Analysis

The pH of water was measured to determine the level of acidity on a regular basis. The electrical conductivity was measured to determine the total of all salts and acids in the water samples and the ability of water to resist changes in pH. This report provides results from a 10-year study from 1965 to 1975 of a watershed in southern Norway. The site is a coastal watershed near Risor, about 230 km from Oslo. The study was quite thorough in examining the weather, water quality, vegetation, and the organisms living in the water and sediment. The chemical analysis provides information on the water quality when the samples were taken. The analysis of the organisms living in the water provides a long-term look at the overall health of the water and the impact on the structure of the food webs present.

The study area included three different regions based on the nature of the underlying rock and sediment. Area A has easily weathered bedrock and sediments consisting of sand and mollusk shells. The shells provide a source of calcium carbonate that serves as a buffer to reduce the impact of acid. Areas B and C have two different types of slow-weathering bedrock, with Area C having the most resistant rock.

The study area was subject to 2–4 g/m^2 of sulfate deposition annually. The prevailing winds in the area are from the southwest and west. The United Kingdom is the closest country in the west and southwest direction. The degree of acidification of water by acid rain is related to the amount of bicarbonate ion in solution, which was measured. The sulfate concentration was also measured. Sulfate can come from acid rain, though some may come from the action of water on soil. The presence of calcium ions, also measured, gives an indication of how much soil has been dissolved by the water.

Acid deposition in the study area had two annual peaks. The first occurred during spring thaw, when the acidic snow melted and sent a spike of acid into the streams and lakes. Some sensitive animals can avoid this spike because the water during this time is stratified: fish can swim to areas of lower acidity and avoid damage. The second spike occurred as the result of heavy autumn rain.

During this period, the water in lakes is well mixed, and organisms cannot avoid the acid.

Lakes and streams in the entire area were characterized by low buffering capacity (ability to neutralize acid) and low dissolved salts shown by low conductivity. The normal buffering of water was not present due to the nature of the underlying bedrock. During periods of acidification, the pH of rivers decreased from 5.9 to 5.4. Similar declines were observed in lakes in the area.

The responses of the three areas to acidic precipitation were very different. Bodies of water in Area A were the best buffered, and they showed the normal seasonal variations of pH during spring thaw and autumn rains. Waters in Area B were affected primarily by a decrease in buffer capacity during the spring and autumn acid increases. Area C, unlike the other two areas, saw a decrease in pH of about 0.5 pH units and a corresponding increase in sulfate.

Area C showed the greatest damage. This was an area in which few people lived, which had negligible human impact during the study period. In the 1950s and early 1960s, there was a significant fish kill in the region after the cutting of the coniferous trees in the area. It is known that tree cutting also releases acid into local waters, and this was the case in rea C. However, the fish kills and other changes observed during this study cannot be related to forestry or land use because the area was not actively managed or farmed during the study.

Biological Changes

The most acidic lakes no longer had fish populations. Acid-sensitive species such as *Daphnia,* leeches, most mayflies, and mollusks were also absent. The species present were filamentous algae and *Heterocope saliens,* a zooplankton. In healthy lakes, fish are the primary predators. In the absence of fish, the primary predators shift to zooplankton. The loss of leeches from the acidic lakes may be due to the lack of the mollusks they feed on. This is an example of how the loss of one species leads to the loss of others.

Each lake in the study area shows a unique combination of organisms, so no generalizations can be made other than that the degree of acidification produced the expected loss of acid-sensitive fish and other organisms, leading to the overgrowth of a new ecology of acid-resistant species. The waters of acidic lakes were also more transparent and less colored due to the absence of microscopic organisms that reduce transparency.

I. Th. Rosenqvist, "Alternative Sources for Acidification of River Water in Norway," *Science of the Total Environment* 10, no. 1 (1978): 39–49. This article questioned the direct link between acid precipitation and the decline of fish stocks. The author pointed to the impact of changes in land use on the chemistry of aquatic systems. During the last century, land in the upland areas that was previously used for grazing has been abandoned for use as farms and was allowed to revert to evergreen forests. These would be expected to produce soil acidification. This study did not find a consistent pattern of acidification of lakes that could be attributed to changes in land use. However, a study in central Scotland found a clear indication that planting evergreens increases stream acidity. The report finds that the lack of forests may increase the impact of precipitation on acidification. The relative importance of different sources of acidification on fish stocks remains unclear.

The eight-year project confirmed that acid precipitation reduces fish stocks. However, the effect of acid precipitation on forest productivity is not clear. To restore lakes to healthy conditions, the project predicted that the hydrogen and sulfate ions need to be reduced by 70 percent. This can only be accomplished by international efforts, and there are major political problems that stand in the way of such sweeping changes in energy policy.

Efforts to restore lakes by adding lime to decrease the acidity and restocking the fish have not worked well. The other possible solution is to develop varieties of salmon and trout that are more tolerant of acidic conditions.

Additional Reference Material

F. E. Ireland and D. J. Bryce, "Pathways of Pollutants in the Atmosphere—The Philosophy of Control of Air Pollution in the United Kingdom," *Philosophical Transactions of the Royal Society of London, Series A: Mathematical, Physical and Engineering Sciences*, 290, no. 1376 (1979): 625–37.

HELSINKI CONFERENCE NEWS

Articles from *Nature* are summarized to provide context and information for the game.

W. Walgate, "Allies for the UK Government," *Nature* 310 (1984): 535.

Three reports just released have provided the U.K. United Kingdom government with considerable support in the negotiations on acid rain over whether the twenty major power plants in the United Kingdom should be retrofitted with scrubbers. All three reports emphasize the complexity of the issue and the uncertainties in the science; none offer quick solutions.

The Watt Committee on Energy suggested that liming of affected lakes was a more cost-effective solution than reducing sulfur emissions. The committee noted that the formation of sulfuric acid from sulfur dioxide requires oxidation in the atmosphere. The major source of these oxidants, according to the Watt Committee, is emissions of nitrogen oxides and hydrocarbons that then form photochemical oxidants such as ozone. They suggest that reducing NO_x might be more effective than reducing sulfur dioxide in reducing the acidity of rain. They also argued that reducing local emissions would be more cost effective for Norway and Sweden than reducing British emissions. They note that to reduce pollution by 1 ton in Norway would require a reduction of 46 tons in Britain. On the other hand, Norway would only need to reduce emissions by 2 tons to get the same effect. Therefore, it would be more cost effective for the United Kingdom to pay for reductions in Norway than in the United Kingdom to achieve the same result. The Watt Committee also argued that the large uncertainties make it unclear how best to proceed. They suggest not spending large sums on the problem before a clear solution is determined.

The House of Lords took a strong position against the current European Economic Community (EEC) proposals for acid rain reduction, which would involve a 60 percent reduction of sulfur dioxide and a 40 percent reduction of nitrogen oxides and particulates by 1995. The House of Lords challenged the cost and benefit estimates of the EEC proposal, which would require scrubbers on twelve of the United Kingdom's twenty large power plants. The House of Lords argued that the EEC underestimated the cost by a factor of 5 and that the true cost would be about $4 billion in capital costs and $800 million a year in operating cost. They also noted that the EEC did not consider the environmental damage resulting from mining 4 million tons of limestone each year to operate the scrubbers nor the disposal of 7.5 million tons of dirty calcium sulfate waste each year. The House of Lords does agree that some control measures are needed and has proposed adding scrubbers to two power plants and a long-term transition to fluidized bed coal combustion technology, which is more efficient and also produces less pollution. These changes would produce a 30 percent reduction in sulfur dioxide emissions by 2000.

The third report by Sir John Mason published by the Royal Society noted that the level of acidic compounds observed was not clearly related to the quantity of sulfur and nitrogen oxide emissions. The report argued that the limiting factor in formation of acidic compounds was related to oxidants in the air instead.

V. Rich, "Polish Pollution: Smelter Shuts Down," *Nature* 289 (1981): 112.

The Ecological Society founded in September in Poland scored a major victory in Krakow. The government has promised to shut the Skawina aluminum factory. Because Krakow lies in a depression that gets little natural air circulation, it suffers from extreme air pollution, including an estimated 70,000 tons of sulfur dioxide from the steel mills in Nowa Huta and 35,000 tons of sulfur dioxide from Skawina, which is powered by a coal-fired power plant.

There is little published information on the problem, but people could see the smoke and dust. Locals worried about the safety of vegetables grown in their gardens, and forestry experts in the nearby Royal Forest had recommended spraying the forest with lime to counteract damage.

The problem became more obvious in 1979 when gold art works at the Wawel Castle museum were found to be corroding. In 1980, the Ecological Society was formed by the Polish government to centralize the local groups protesting for environmental protection. The Environmental Society demanded that the government provide accurate information on all emissions, including fluorine and hydrogen fluoride from the aluminum plant. They also demanded that the aluminum process be shut down until the hazards of these toxic emissions could be eliminated. The government had already planned to modernize the two mills at a cost of $4.5 million for the aluminum plant and $300 million for the steel mill.

In December, the debate spread to the national television network, and the government ordered thirty-five obsolete aluminum electrolysis tubs to be taken out of service. The mayor of Krakow increased that to 160 tubs, which cut emissions by 60 percent and production by 50 percent. By the end of the first week of January, the decision was made to completely close the Skawina plant and transfer aluminum production elsewhere in Poland.

J. N. B. Bell, "Acid Precipitation—a New Study from Norway," *Nature* 292, no. 5820 (1981): 199–200. Public concern about the destruction of freshwater fish stock due to the increased acidity of lakes and streams has been strongest in Norway. In the mountainous regions of southern Norway there has been an accelerating decline of fish stocks, with extinction of fish populations in an area of 13,000 km^2 and severe problems over another 20,000 km^2.

To better study the problem, a major Norwegian interdisciplinary research program was established. The research was conducted from 1972 to 1980, and the study's findings have recently been summarized in their final report.

The final report made a major contribution to the understanding of the relationships between the chemistry of precipitation, soils, and aquatic ecosystems. A major problem lies in the difficulty of projecting long-term effects from short-term studies of acid precipitation, given that the changes in soil and water chemistry being measured are relatively small.

In spite of these problems, the fact that the precipitation in southern Scandinavia has become more acidic due to the transport of pollutants over long distances has been established beyond doubt. Precipitation at many sites was sampled for concentrations of hydrogen ions, sulfate ions, and nitrate ions to determine the amount of pollution from rain. The importance of dry depositions directly onto vegetation and soils remains uncertain, so it is not known whether this makes a large contribution to total acid deposition in some areas.

The report concedes that there are few reliable records of the acidity of lake waters before 1950, but the marked increase in acidity since then has been well documented.

This project assessed the input of specific ions in annual precipitation and the output of these ions into waterways. It appears that the mobilization of aluminum ions from soils is a major factor in the loss of fish stocks. The combination of acidic conditions and the presence of aluminum ions interfere with the ability of fish to absorb salt into their gills. The absorption of salt is necessary to maintain proper fluid balance.

The problems are particularly severe when the precipitation is in the form of snow. This is especially problematic in southern Norway where snow makes up a large proportion of the precipitation. Pollutants build up in the snowpack during the winter and then are released very quickly into lakes and streams during the spring when the snow melts. This results in a very rapid increase in acidity in waters, sometimes referred to as an "acid flush." The project has examined the chemicals in melt-water and found that up to 80 percent of the pollutants are released in the first 30 percent of the melt-water. This is especially damaging to the highly sensitive larval stages of salmon and trout.

Studies on the effects of acid rain on plants and trees have not provided any definitive support for predictions of serious damage to the timber industry. This is partly because the relationship between acidic conditions and the uptake of nutrients is complex. The report implies that there must be reduced tree growth due to acid precipitation but offers little evidence in support.

J. Becker, "Swedes Persist," Nature 295 (1982): 641.
Sweden plans to hold a two-week conference on acid rain to persuade its European neighbors to take the problem seriously. The conference to be held in Stockholm will include fifteen environmental ministers from East and West to examine the scientific evidence, followed by a political debate. The sole country not being represented is Britain, the major contributor to the transnational problem of acid rain.

At a recent meeting in Brussels organized by the European Environmental Bureau, many experts admitted that no one was entirely clear about the precise mechanisms of cause and effect of acid rain. There was evidence to show that over the past two years dramatic changes have taken place in the forests and lakes of Scandinavia. Research is also available to show that sulfur dioxide emissions may also be linked to acidity in the freshwater of Belgium and damage to forests in Germany. The environmentalists argue that the problem is not just a Scandinavian one and is already too serious to wait for irrefutable scientific evidence. Much of the pollution is coming from Eastern Europe, notably East Germany and Czechoslovakia. The poor economic and energy problems of some of these Eastern European countries makes it highly unlikely that they would consider making capital investments in costly antipollution equipment.

Although thirty-four countries have initialed the Geneva Conference protocols, Greece, Belgium, and Italy have failed to ratify it.

Sweden's attempts to create the political climate to initiate change include the UN Conference in Stockholm in 1972, the Helsinki meeting in 1975, and the 1981 Conference in Gothenburg. Despite some of the evidence presented by a study by the Organisation for Economic Co-operation and Development, the European Commission still resists the idea of the need to do anything about the problem. Many of the member nations such as the United Kingdom still remain openly skeptical about the cause and effect of acid rain.

J. Becker, "Acid Rain: UK Unrepentant," Nature 298, no. 5870 (1982): 112.
A modest success for the Swedish government was scored last week by convening a meeting of the 1979 Geneva Convention signatories to accept a report on the state and knowledge about acid rain. This means that the convention will come into force by the end of the year.

The Geneva Convention on long-range transboundary air pollution established that polluting countries must reduce sulfur emissions, after accepting the report's results that the damage done to about to 20,000 Scandinavian lakes and a million hectares of central Europe's forest is without a doubt due to sulfur dioxide and nitrogen oxide emissions.

The worst polluters, the United Kingdom, France, and the United States were rather complacent about this. Britain's Under-Secretary of State for the Environment, Giles Shaw, admitted that the United Kingdom was the biggest polluter but maintained that the United Kingdom has already reduced its emissions by more than 20 percent. This was mainly due to economic recession, the use of natural gas and low-sulfur oil, and a decreased use of coal, rather than any direct action to reduce emissions.

The big surprise at the conference was that Chancellor Helmut Schmidt of West Germany completely changed his stance on reducing emissions due to his desire to win back the support of the citizens concerned with ecology (known as the Green Party) after his party's near defeat in the Hamburg elections.

T. Beardsley, "Acid Rain: What Cost, What Benefit?," *Nature* 306, no. 5944 (1983): 363.

The Department of the Environment of the British Government agrees that acid rain has contributed to the acidification of lakes and streams in sensitive areas. However, they assert that there is no conclusive evidence to show that the long-range transport of acid pollutants is responsible for the damage seen in the forests in West Germany.

The British government to date has spent 1 million British pounds and plans to spend more on research this year.

Some of these problems were aired at a previous meeting in London of the Watt Committee on Energy. Sir John Mason, who is director of the study for the Royal Society in collaboration with the Norwegian Academy of Science and Letters and the Royal Swedish Academy, feels that it will be hard to know what the actual results represent in terms of sources and effects. The study will concentrate on the chemistry of rainwater as it percolates through soil.

There are a few theories on how acid rain kills trees. It may turn out that a large financial investment to reduce emissions from coal-fired plants has little benefit; further research seems to be the best option. While nothing was settled, the British government agreed that a 30 percent reduction of emissions would cost about 1,000 million British pounds.

The governments of Norway, Sweden, and West Germany take a more positive stance. West Germany has introduced measures to reduce the emissions of sulfur dioxide and nitrogen oxides. Norway and Sweden are distributing pamphlets to tourists stating that their lakes are as clean as the British Energy Industry.

The European Community is attempting to control acid emissions by proposals that all member states should reduce their emissions by some proportion by 1995. One proposal calls for a 50 to 60 percent reduction in emissions.

"Acid Rain, Political Bile: The Lack of Scientific Solution Should Temper the Political Response to Acid Rain in Europe," *Nature* 303, no. 5920 (1983): 739–40.

"The most complicated issues in science have a habit of being the most political." The issue being considered is acid rain as well as the mix of uncertain science and political posturing. The tension between politics and science is played out in determining solutions to the acid rain problem.

The West Germans and Scandinavians want the polluter to pay, yet the polluter may be further west in Europe. There is pressure on German and Scandinavian scientists to provide evidence of the bad effects of acid rain, and the British and the French scientists are under pressure to show the reverse. All scientists need to avoid being pressured and should let the scientific evidence speak for itself. Unfortunately, there is not enough time to collect definitive evidence. Decisions must be made before the bulk of the scientific evidence is in, and this is where politics come into play.

Scientific understanding of acid rain is still under investigation, and at this point there are more questions than answers. If scientists are forced to make a decision, external factors such as personality and cultural background will come into play.

The politicians' point of view is that they will be provided with more or less biased advice, and there will be much doubt. When science is working properly, doubts diminish with time as more data are obtained. Therefore, an interim decision on emission levels for airborne pollutants with modification as needed seems like a reasonable solution.

A draft directive is proposed that puts emission limits on factories and was approved by the Council of Ministers within a record time of two months. This is a small step. Agreement on the next steps, which will involve setting limits on levels of emissions, will be much harder to accomplish. But this is an evolving science, so adjustments to the sharing of risk and cost can be made later.

R. Walgate, "Royal Society Appointed Referee in UK Dispute," *Nature* 304 (1983): 85.

The Royal Society of London in conjunction with the Norwegian Academy of Science and Letters and the Royal Swedish Academy have begun a five-year study of acid rain in northern Europe. This is groundbreaking for society because this is the first formal study of controversial public health issues. The study is unprecedented in its high cost: 5 million British pounds. The study is being funded by the two British nationalized industries that are mostly likely to be negatively affected by its results—that sulfur dioxide emissions from British power stations are creating acidic conditions and thus are responsible for the disappearance of fish from lakes in Scandinavia.

The study was proposed by Sir Walter Marshall, fellow of the Royal Society and now chairman of the Central Electricity Generating Board, who feels that based on current evidence, reducing emissions by the use of scrubbers at the cost of 4,000 million British pounds is not a reasonable solution.

All the parties involved wish to emphasize that the study of acid rain will be carried out independently of both the sponsors and the governments. Members of the management group for this study include Professor T. R. R. Southwood (Oxford), who is the present chairman of the Royal Commission on Environmental Pollution, as well as Dr. T. F. West, Professor E. J. Denton, and Dr. J. F. Palling, all from the Freshwater Biological Association.

This study differs from other ongoing studies in that it focuses on the effects of acid rain on land and water rather than the airborne transport of sulfur dioxide from one place to another.

The Central Electricity Generating Board on its own behalf and that of the National Coal Board refers to an earlier study in 1976 that suggested that Britain might be responsible for about half of the acid rain in Norway and Sweden.

By their approval to engage in research on this subject, the two British boards signal that they accept responsibility for doing what they can to lessen problems that they may have caused. By commissioning the Royal Society to perform the study, they have for practical purposes appointed it to be the referee in deciding what can be done to lessen the problems caused by acid rain. Five years from now, it will be interesting to see how well they have done in that role.

R. Walgate, "Acid Rain Research: Too Late for Black Forest?," *Nature* 303, no. 5920 (1983): 742.

A consortium of research groups in Munich has been formed to study how acid rain is destroying the Black Forest, a large stand of fir and spruce trees that covers the upper Rhine and is one of the major recreational areas in Europe. A number of research scientists are alarmed by how fast the trees are deteriorating and how little is being done to find ways to stop the damage.

A symptom of this deterioration is that honey production, which is a good gauge for the overall health of the forest, has failed catastrophically in the Black Forest. Old trees, those that are over a 100 years old, are losing their needles. Young trees near the tree line are also dying where they are exposed to nitrogen oxides in a mix with sulfur dioxides. These compounds are cited as the main reason for the destruction of the trees. Other minor reasons may be the growth of acid-loving fungi, which can affect root hairs.

The question is, what can be done to prevent further destruction? Other than to investigate how acid rain leads to the destruction, the key step is to control the emissions of nitrogen and sulfur oxides. This must be done on a Europe-wide agreement. Reaching an agreement may take much longer than the research needed to understand the science, and at that point it would be too late.

V. Rich, "Czechoslovak Pollution: Past Sins Now Whitewashed," *Nature* 303, no. 5915 (1983): 276.

The Party daily newspaper, *Rude Pravo,* recently reported the fact that almost one-third of Czech forests are damaged by pollution and this threatens the nation's important timber industry. Five years ago, it was not possible to publicly discuss such issues. The loss of trees ranges from 10 to 70 percent over the affected area, but the hardest hit forests have required total clearing and replanting. In the Krusne Mountains, the forests were replanted with trees thought to be resistant to pollution, but the strategy was unsuccessful due to the extreme levels of pollution.

The annual loss of timber production has been as much as 8 percent of the total annual production, which the government first acknowledged last year. The forest damage has been attributed to the use of lignite to fuel Czech power plants, and further damage has been caused by insects that attack the trees weakened by pollution.

Following the normal practice of the socialist government, the problem was acknowledged only when a solution could be announced as well. The government reported that a catalyst had been developed that would reduce nitrogen oxide emissions by 90 percent. The technology has already been installed on one power plant to remove 4,000 tons of nitrogen oxides each year. Unfortunately, nothing was said about any efforts to reduce sulfur dioxide emissions.

K. Mellanby, "Ecology: Acid Precipitation and the Black Forest," *Nature* 304, no. 5926 (1983): 486.

Kenneth Mellanby is editor of *Environmental Pollution* and also chairperson of the Watt Committee on Energy working group on acid rain.

While there has been a great deal written in newspapers and shown on TV about the damage that acid rain has had on trees in the Black Forest, there is no real scientific evidence to support this. At an international conference where more than fifty papers were presented on this topic, there was very little evidence to support this theory.

At this meeting, topics that were presented included methods used to measure the emission of sulfur and nitrogen oxides, mechanisms to explain how these oxides are converted to acids, and explanations for how these acids wind up in soil, plants, and water.

The presentations showed that the decline in the forests in West Germany is not as widespread or alarming as had been predicted. However, there is still concern that fir and spruce are not growing at the usual rate. Although this effect has been mostly confined to one area, it is hard not to worry about this spreading quickly to other parts of the forest.

There was no real agreement about why this reduction in tree growth is happening. It was suggested that perhaps a cold winter followed by a dry summer, fungal disease, or even increasing ozone levels with an increase in aluminum due to acid rain were all contributing causes. B. Ulrich (Institute of Soil Science and Forest Nutrition, Gottengen) tried to make a strong case for acid rain leading to an increase in aluminum levels as the main reason for the damage seen. There was little support for his position. Weakening his position further were reports from Scandinavia that the predictions of major forest destruction have been greatly overestimated.

There is a consensus that in the past thirty years lakes and rivers have become more acidic and a large number of fish have died. There is no general consensus on the cause of these deaths. I. T.

Rosenquist at the University of Oslo did some work using ion-exchange methods in soil and water and found that an increase in acidity may be due to changes in land-use rather than the acidity of the precipitation. He showed that increases in acidity had occurred long before the emissions of sulfur and nitrogen oxides rose to their present levels.

K. Melanby writes that, in his opinion, some of the presenters at this conference did not distinguish between the effects of dry deposition such as dust and particulates that settle out of the atmosphere and wet precipitation such as snow and rain. Another speaker, S. Luckat from the University of Dortmund, was able to demonstrate that dry precipitation of sulfur dioxide on structures such as stonework and statues will cause serious damage, but acid rain does not contribute to this sort of damage. The rain may initially have a high concentration of acid, but successive rainfall is less acidic and serves to wash the statues clean.

Mellanby took an extensive tour of the Black Forest and was surprised to find that there was little evidence of serious damage. There were some damaged trees, but the guide conducting the tour was not able to show a single tree that had died. The damaged trees were found in areas at high altitudes and thus may be more stressed than trees at lower altitudes.

Mellanby concludes with the thesis that damage in the Black Forest is not likely due to sulfur dioxide emissions in the Federal German Republic. Sulfur dioxide concentrations in the forest are relatively low, and there is an abundant quantity of lichens growing on the foliage, leafy plants, and even some of the damaged trees. If damage was caused by acid rain in the form of sulfuric acid, then it must have occurred a long time ago. Mellanby thinks that the policy of the West German government to reduce sulfur output from industry to very low levels is being driven by the political clout of the Green Party. Reduction to very low levels is going to be very expensive and ultimately will not have any effect (good or bad) on the trees in the Black Forest.

V. Rich, "Hungarian Environment: Belated Help for Teachers," *Nature* 304, no. 5925 (1983): 387.

Hungary's National Authority for Environmental Protection and Nature Conservation has announced that environmental education will be added to teacher training this fall. It is surprising that it has taken more than a decade to do this as environmental education was added to primary and secondary education in 1972. Furthermore, a 1976 law called for all citizens to be made aware of environmental issues in many different venues. Environmental education has only recently been added to the technical school curriculum.

The people of Hungary have been somewhat aware of visible environmental problems such as the decline in water quality at the resort area of Lake Balaton. Actions have been taken to reduce the input of fertilizer and pesticides into the lake. However, people do not seem to be aware of pollution that they cannot see or smell, such as acid rain.

J. S. Dunnett, "West German Ecology: Ozone Named as Culprit," *Nature* 301, no. 5898 (1983): 275.

West German forests are some of the best managed forests in the world. Flying across West Germany, the forests look like the patches of crops they are. However, the damage to spruce trees first became obvious in 1976 in the Black Forest and in Bavaria. Fir trees have also begun to show damage, and in 1982 the damage had spread to other species and other areas of Germany. Damage was most pronounced at higher altitudes and on the southwest-facing forests.

Until recently, the damage had been attributed to acid rain, possibly due to root damage. However, the Pollution Control Authority of the state of North Rhein-Westphalia has released an interim report on their recent study that concludes the actual problem is ozone. The study found that exposing trees in the laboratory to rain of pH 4.5 and 3.0 caused little damage. On the other hand, exposure to 100 to 400 $\mu g/m^3$ of ozone caused clear damage to needles. They also noted that lichens were quite sensitive to pH but not to ozone. This means that damage to lichens and trees when viewed together provides an indicator of the cause of the damage. Most areas of the Black Forest have strong growth of lichens but damaged trees.

The most heavily damaged forests in West Germany are on the border with East Germany and Czechoslovakia, where the damage is as high as 40 percent. West Germany cannot do anything about this imported pollution. East Germany gets 70 percent of its energy from brown coal, using 270 million tons in 1982. Of course, West Germany also exports a large amount of pollution, but this is being reduced by a program to install scrubbers on most coal-fired power plants by 1990.

Pollution has had a significant effect on German politics and led to the dissolution of the Social Democratic Party (SPD)–Free Democratic Party (FDP) coalition government last year. The FDP is now in a coalition with the Christian Democratic Union, and stronger environmental voices have left the government. FDP leaders who oppose environmental controls are being supported by the leadership of the Christian Socialist Union from Bavaria, who continue to downplay obvious environmental damage. If the SPD wins the March elections as predicted, it is expected to establish a Ministry for the Environment to unify environmental protection under a single authority.

"Acid Rain: Parliament in Power Row," *Nature* 311, no. 5987 (1984): 597.

The House of Commons Select Committee on Energy issued a report that is a direct attack on the Central Electricity Generating Board (CEGB) and differs significantly from the recent report issued by the House of Lords. The House of Commons report accused the CEGB of "ignorance or a deliberate attempt to mislead." The CEGB has argued that damage to German forests is not due to sulfur dioxide but to ozone and nitrogen oxides. Some feel that the House of Commons report does not give adequate credit to the CEGB for its research program and efforts to find cost-effective pollution control technologies.

Sir Walter Marshall, head of the CEGB, countered that "if reduced CEGB sulfur dioxide emissions would solve real environmental problems in a cost effective way, we will get on with the job despite the costs." He estimates the cost to retrofit a power plant to be nearly $300 million. (The CEGB operates 20 such plants.) He also noted that the problem of controlling nitrogen oxides is more difficult to solve than for sulfur dioxide.

The CEGB will be experimenting with new burner designs at two plants that would be more efficient and thus reduce emissions of sulfur dioxide. However, Marshall noted that both designs might actually increase emissions of nitrogen oxides.

T. Beardsley, "Acid Rain: Uncertainty Persists in Europe," *Nature* 307 (1984): 101.

A report prepared for the U.K. Department of the Environment has demonstrated the serious lack of information and sampling sites. It was impossible to document long-term trends in acid, nitrate, or sulfate. The report recommended funding a much larger network of wet-deposition collectors.

The general trends observed where data were available were that the deposition of sulfate had increased since 1900 and nitrate since 1957 corresponding to increases in emissions. Where good data were available it showed that 70 percent of acidity was due to sulfuric acid and 30 percent from nitric acid. Cumbria and some areas of Scotland had the highest levels of pollution, and these areas were comparable to polluted areas in Scandinavia and North America. They also noted that acidity is not evenly distributed over the year. In some places, up to 30 percent of the year's acid deposition occurred in a single week.

The EEC Council of Ministers has not been able to agree on a plan to reduce pollution. One article in the proposed plan to establish emission standards for the entire EEC was unacceptable to the United Kingdom and Ireland. A proposal to drop the EEC-wide standard was opposed by the Netherlands. A second proposal to be debated soon is a more limited plan that applies to old and new power plants and allows countries flexibility in how to meet the standards. This proposal is also expected to be controversial.

D. H. P. Laxen, "Linear Scale for Acid Rain?,"
Nature 309 (1984): 26.

In the ongoing public debate about acid rain, it is important to carefully represent the results of acidity measurements of rainfall and freshwater in the most appropriate manner.

Acidity of rainfall and freshwater is commonly reported in terms of pH. The pH scale is a convenient way to express the concentration of hydrogen ions (H^+) over many orders of magnitude. Difficulties may arise from a failure to realize that pH is a logarithmic function rather than a linear one. Confusion in circles of both scientists and nonscientists can easily arise. A change in pH of 5.0 to 6.0 may not seem significant but actually represents a tenfold change in concentration of hydrogen ions.

More specifically, definition of the trends over time in acidity levels has garnered a great deal of attention. This is important in studying the relationship between emissions of sulfur dioxide and nitrogen oxides on acidification of the environment. Unfortunately, pH records of rainfall and natural water are only available for the past several decades, and there are problems with the continuity and quality of the data. Another method, diatom stratigraphy in lake cores, has been used to reconstruct the pH history of a lake in the southwest of Scotland over the last 400 years or so. *Diatoms* are a major group of algae, and they are among the most common types of phytoplankton. *Diatoms* are unicellular, although they can form colonies in the shape of filaments or ribbons. This method takes a core of lake sediment and examines the fossil shells of diatoms in different layers. The layers can be dated accurately using the radioactive decay of lead isotopes. The types of diatoms that live in different pH values are well known, and the distribution of diatoms in each layer of sediment allows the pH of the lake at the time to be determined.

Figure 5 shows the pH history of Round Loch of Glenhead using a logarithmic (pH) scale, which is compared with the linear scale shown in Figure 6.

Note that the results in Figure 5 indicate that about half the total change in pH took place by around 1900. It appears that almost all the change had taken place by 1930, with very little change seen after 1940. This figure suggests that the major pH change occurred early in the first half of the twentieth century. Figure 6, however, indicates that half the pH change had taken place by the later date of 1920, and there was a further steep increase from around 1950 to the present.

The use of a linear representation of concentration provides the most appropriate means of representing the magnitude of change in acidity, and there seems to be a move to use this in reporting rainfall chemistry data, a move that is long overdue.

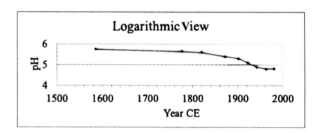

FIGURE 5 The pH history in Round Loch of Glenhead using a logarithmic (pH) scale. (Data from Laxen, 1984.)

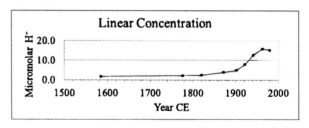

FIGURE 6 The pH history in Round Loch of Glenhead using a linear concentration scale (micromolar). (Data from Laxen, 1984.)

"West Germany Takes the Lead," *Nature* 307 (1984): 306.

A government committee recently reported that the serious damage to German forests cannot be explained by any normal changes and must be at least partially due to air pollution. They report that the success rate for survival of new tree planting is the worst in the world. In response, the German Legislature, the Bundestag, has passed regulations to reduce these emissions. A limit of 1,800 mg sulfur/m^3 has been imposed for new power plants, and the limit for old plants has been set at 3,600 mg sulfur/m^3. The goal is to reduce the 43 percent of acid emissions due to power plants. This reduction is an improvement, but the West German environmental council hopes to see a reduction in line with what is done in Japan, where the limits are 100 to 500 mg sulfur/m^3. Even under the current weaker limits, West German emissions would fall from the present 3.5 million tons to 0.55 million tons by 1995. This reduction may be hard to achieve due to the recent opening of the largest brown coal mine in the world by Rheinischen Braunkohlen Werke AG, which is expected to produce 2.5 billion tons of high sulfur coal over the next 50 years.

The federal government is also spending almost $20 million each year on research on the problem, compared with only about $1 million being spent in the United Kingdom. German states are also studying the problem. Baden-Wurttemberg has built three pilot plants to investigate techniques for removing sulfur, at a cost of over $20 million, without support from either the federal government or the EEC.

Nitrogen oxides from car exhausts are another issue, and some have proposed reducing the speed limits on all roads and the autobahn as a way to reduce these emissions.

T. Beardsley, "UK Inspector Protests at Brussels," *Nature* 307 (1984): 24.

The Chief Alkali Inspector Dr. Leslie Reed strongly opposed two laws proposed by the EEC for pollution control. Dr. Reed pointed out that the 1863 Alkali Act already requires that plants in the United Kingdom employ the "best practicable means" to control their pollution and that the Inspectorate sets emission standards. He does not believe a second layer of regulation is needed. The current U.K. law allows for rapid response and adaptation to local circumstances.

Britain has objected to the EEC limits, which passed with a two-thirds majority, and is expected to oppose a second law that would set limits for total national emissions. West Germany is the major driving force in this proposal due to the recent severe damage to German forests.

Dr. Reed also objected to a third proposal for reducing nitrogen oxide emissions. He objects to any standards that are expensive and not based on clear evidence that they are needed. He objects to laws for Britain that benefit "one or two European countries which have inadequate controls of their own."

Mr. Ian MacGregor, chairman of the National Coal Board, also objected to the proposals. He commented that "after we cripple some of our basic industries, we could then find we have not solved the problem." MacGregor noted that volcanic activity is a major source of sulfur dioxide. A second argument is that the forests suffering damage to evergreens have healthy lichen growth. It is well known that lichens are very easily damaged by sulfuric acid pollution. Finally, he argued that liming (using a basic substance to neutralize acid) was the best approach until there is conclusive scientific evidence that more expensive measures will solve the problem.

"CEGB Takes a Pasting from MPs," *Nature* 311 (1984): 94.

An all-party committee of Members of Parliament (MPs) published a report that stated that Britain should drastically cut back on sulfur dioxide and other emissions that contribute to the problem of acid rain. Most of the burden of this recommendation would fall on the Central Electricity Generating Board (CEGB). If the MPs recommendations were put into effect, it would require the CEGB to cut emissions by 85 percent. This would be very costly and would raise electricity bills by 10 percent. Furthermore, according to the CEGB, there is no guarantee that the measures taken would have any effect on acid rain itself.

But the MPs counter by saying that after many months of obtaining evidence from politicians and scientists in Britain, West Germany, and Scandinavia they are convinced that sulfur emissions are damaging to the environment and action must be taken immediately rather than waiting for all of the evidence to be in. In this they agree with a report submitted by a committee from the House of Lords (W. Walgate, "Allies for the UK Government," *Nature* 310 [1984]: 535).

The MPs impressed by the available scientific evidence linking sulfate emissions to acid rain and "appalled" by lack of action in Britain, state that Britain must (among other measures)

- Join the "30 percent club" of nations committed to the 30 percent reduction in total national emissions of sulfur dioxide by 1990, a target to be achieved solely by cutbacks on CEGB emissions.
- Refit CEGB power stations to achieve a national 60 percent reduction in sulfur dioxide emissions by 1995.

The Commission of the European Communities (CEC) would like to see the measure applied to all plants of output greater than 50 megawatts. The British MPs have singled out electricity supply and would like to see the CEGB reduce emissions of sulfur dioxide by 85 percent, to achieve the net national average of 60 percent reduction.

The MPs claims that this is reasonable to ask of the CEGB because their emissions have remained constant while net British emissions have fallen 37 percent.

Although it may seem that the MPs are singling out the CEGB for punishment, this is deserved because the CEGB has been dismissing the effects of acid rain. The CEGB claims that to dissolve one inch of stonework by acid rain it would take 5,000 to 20,000 years. The MPs say that it is well understood that dry deposition leads to the much more rapid process of flaking. In this context, the report quotes the cathedral architect at Cologne saying that none of the original stonework of the cathedral will be there in five years. The MPs report that the same process is at work in Britain.

In other areas the CEGB has made no attempt to monitor the effects of acid rain and is spending greater sums of monies to study the problem than the monies spent on research on desulfurization.

As for the scientific evidence, the MPs feel that it took pains to get a balanced and detailed view in the United Kingdom and abroad. The House of Lords also performed a study, but according to Sir Hugh Rossi, the chairperson of the committee writing the report, they did not do their homework in that they did not go abroad and their report was consequentially very superficial.

The work of the U.K. Natural Environmental Research Council (NERC) scientists impressed the MPs because it was independent of the interests of manufacturing or industry and indicated that there was enough evidence for a decision to be made now rather than waiting five more years and doing nothing.

Further, the suggested solution of adding lime to acidified lakes in Scandinavia ignores both Scandinavian research and mere geography and makes liming (even as a temporary measure) highly impractical.

The only solution is for Britain to put its own house in order say the MPs. Will the MPs advice be taken? The next step will be a debate on the report.

It should be noted that Britain's Social Democratic Party (SDP) announced their own solution, which is energy conservation. They claim that if the government invested more in energy efficiency measures they could achieve reductions in sulfur dioxide emissions without the costly installation of desulfurization equipment in power stations. A 5 percent cut in electricity generation would enable

Britain to meet the Select Committee's targets by 1990, at half the cost estimated by the CEGB. This would also improve the competitiveness of British industry.

V. Rich, "Immediate Action Imperative," *Nature* 308 (1985): 501.

Poland is facing a possible ecological catastrophe according to Dr. Antoni Kuklinski, head of the Committee of Spatial Development of the Polish Academy of Sciences. Dr. Antoni identified three areas that had reached the "breaking point." These areas are 1) the natural environment, 2) the technical infrastructure which includes transport, power supplies, and water management, and 3) the poor conditions in large urban areas.

A suggested solution is to invest more monies in solving the problems of the environment, even if this means that monies invested in industry are reduced to only 30 percent of the national total. At the same time, the housing sector must learn to make better use of the available resources. According to Dr. Antoni's article in the government newspaper, *Rzeczpospolita,* the whole structure of investment must be changed, with greater powers given to local "self-government" bodies. "Investment in the environment must be of primary concern."

This is not the first time concern about the environment has emerged in the Polish media. Recently, Warsaw Radio noted that in the upper Silesian industrial belt 430 out of 100,000 people die prematurely because of the environmental conditions, and this death rate is 50 percent higher than the national average. Circulatory diseases, cancer, pulmonary diseases, and infant mortality occur much more frequently in this area of Poland. One doctor recommended that people not work in this area for more than 20 years and that those who are not working should live elsewhere.

This view is reinforced by a report that includes the findings of a seminar held in Lublin in the past fall on the chemical threat to the Polish environment. According to a survey carried out in 1982 by the Planning Committee of the Council of Ministers, this area encompassing a population of 11 million (30 percent of the total population) is under severe ecological threat.

Special attention is focused on the classification of all forms of air pollution in terms of sulfur dioxide toxicity equivalent because recent studies have made Polish planners aware of the problem of sulfur dioxide emissions. The hazard from heavy metal dust from the metallurgical and engineering industry is equivalent to 13.8 million tons of sulfur dioxide per year. Gases from the chemical industry have a sulfur dioxide equivalent of 1.26 million tons per year. The annual emission of sulfur dioxide by Polish industry is 2.45 million tons.

The report stresses that the whole pollution problem was studied afresh in 1982 and that actions taken to reform the economy and focus on the three strategic problems of food supplies, housing, and restoration of the ecology will be introduced during plans for 1986 to 1990.

HELSINKI CONFERENCE: COUNTRY-SPECIFIC SCIENTIFIC RESEARCH

Technical papers pertaining to the countries are summarized here for use in the game.

Sweden

G. Jacks, G. Knutsson, L. Maxe, and A. Fylkner, "Effects of Acid Rain on Soil and Groundwater in Sweden," in *Pollutants in Porous Media, Ecological Studies: Analysis and Synthesis*, vol. 47, ed. B. Yaron, G. Dagan, and J. Goldshmid, 94–114 (Berlin: Springer Verlag, 1984).

Jacks et al. provide a broad overview of the geochemistry of the acidification of water and soil for several counties in western and southern Sweden with data on emissions and acidification. This is the area with the most severe acidification because 75 to 80 percent of the acidic pollution in Sweden originates from western and central Europe rather than from local sources. A large local source was also identified in Stenungsund and in Karlshamn that has a pronounced local effect on groundwater.

A time series study of pH, sulfate, nitrate, and hydrogen ions is presented for precipitation in Forshult in Varmland County. Over the period from 1955 to 1980, the pH dropped by over 1 pH unit from an initial value of 5.5 to about 4.3 in 1980. The entire period from 1965 to 1979 had pH values 4.5 or less, meaning that the decrease in pH occurred between 1955 and 1965. Over this period, sulfate increased from 25 ppm to 32 ppm and then varied between 30 and 45 ppm for 1965 to 1979. Nitrate increased even more, rising from 15 ppm in 1955 to 35 ppm in 1979.

Effects on Lakes

The acidification of lakes is particularly severe in southwestern Sweden, where airborne pollution is ten times worse than in northern Sweden because the sources of pollution lie south of Sweden. In addition to direct measurements of the pH of the lakes, the pH levels over the 12,000 years since the last ice age were determined by studying fossils in lake sediment. Before the industrial revolution, the pH of the lakes was around 6.0 compared with the current values of 4.5 for many lakes. This decrease occurred over the past fifty years. There are over 18,000 lakes larger than 10,000 m^2 in western Sweden that are acidified, with pH values of less than 5.5, and 4,000 of these have pH values below 5.0 and are seriously acidified.

The graph in Figure 7 shows the time series constructed to show this change. Figure 8 shows the recent changes for the same lake.

Soil Acidification

The chemical processes related to acidification of soil are complex and depend on a variety of factors, including inputs of acid precipitation, fertilizer, and the nature of the soils. The major factors are

- Wet and dry deposition of acid from pollution
- Nutrient uptake by plants

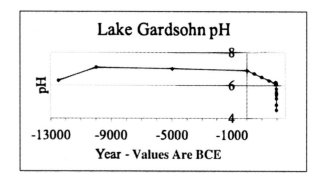

FIGURE 7 Long time scale pH changes in Lake Gardsohn. (Data from Jacks et al., 1984, p. 97.)

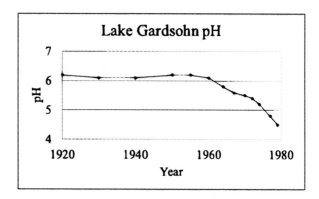

FIGURE 8 Recent changes in pH in Lake Gardsohn. (Data from Jacks et al., 1984.)

- Chemical changes in sulfur and nitrogen compounds due to oxidation
- Chemical changes due to iron content in soils

The natural processes of plant growth and decomposition tend to produce acidification and neutralization, respectively. Therefore, undisturbed ecosystems tend to remain at stable pH values unless they receive acids from external sources. However, when forests and crops are harvested and removed from the land, the decomposition process does not occur, and the general trend is toward greater acidification.

Table 7 shows the relative importance of the three major factors in adding hydrogen ions to soils (decreasing the pH). The data are divided into

TABLE 7 Sources of soil acidification by soil type

	Ecosystem (g hydrogen ion/hectare/year)		
Source	Agricultural Land (Good Soil Quality)	Spruce Forest (Medium Quality Soil)	Pine Forest (Poor Soil)
Acid precipitation	800	1,500	1,200
Nutrient uptake	5,000–10,000	200–700	100
Biomass accumulation in litter	NA	700–1,400	200

Source: Jacks et al., 1984, p. 99.
Note: NA = not applicable.

three different ecosystems. The units of hydrogen ions are grams of hydrogen ions per hectare per year. The data in Table 7 show that acid precipitation is the predominant factor in acidification only for pine forests and is only a minor factor for good-quality agricultural soils.

The forest soils in southwestern Sweden are very thin and are stratified into distinct layers. The underlying bedrock breaks down slowly. The top layer in forests is a litter of needles and leaves, which slowly decomposes from the action of bacteria and fungi. This layer tends to be highly acidic, with pH as low as 4.0 being common. Organic compounds leach from the top layer into the lower layers and cause the release of positively charged ions such as iron, aluminum, calcium, and magnesium from the rock. The components of the underlying rock and sediment that provide neutralization of acid are separated in the lower layers of the soil. Plant roots are the primary way that these ions move from the lower layers of soil to the top layer to neutralize the acid.

The release of aluminum from soil and rock is important in damage to forests and fish. Aluminum is toxic to plants and is the primary toxin that kills fish in acidic lakes. Aluminum takes several forms. It reacts with organic compounds produced in the leaf litter to form substances that do not dissolve in water. Normally, forest streams are brown due to the presence of organic compounds in the runoff. In areas with severe acidification, the organic compounds have been made insoluble, so surprisingly the streams in these areas appear clear. These inorganic aluminum ions, which are not bound to organic material, are the most toxic to fish when they wash into rivers and lakes.

Acidified soils have detrimental consequences:

- Positive ions that are vital plant nutrients—calcium and magnesium—are washed out of the soil, leading to nutrient deficiencies that damage plants.
- Bicarbonate, the natural buffer in water, is decreased.
- Hydrogen ions are free to percolate through the soil along with sulfate ions.
- Toxic inorganic aluminum is mobilized in the lower levels of soil and drains into wells and streams.
- Dissolved organic material in soil and runoff water is decreased (the organic material normally binds to toxic metals and makes them unavailable).
- Heavy metals, zinc, lead, and cadmium are released from the soil and enter lakes and streams.

- Aluminum ions precipitate the nutrient phosphate in soil and runoff so that it is not available to support plant growth.
- The nutrient selenium is bound to the soil and becomes less bioavailable.

Cadmium levels were measured in the soil and in lake water as a function of pH. (Cadmium is one of the most toxic metals to animals and humans.) At pH levels above 6.0 there is essentially no significant cadmium in either soil or water. Between pH 6.0 and pH 5.0, the cadmium concentration begins to rise up to about 0.2 ppm. Below pH 5.0 there is a rapid rise in cadmium concentration. By pH 4.5 the concentration more than doubles, and between pH 5.0 and pH 4.0 there is up to a twentyfold increase in cadmium concentration.

Ireland

B. E. A. Fisher, "Deposition of Sulfur and the Acidity of Precipitation over Ireland," *Atmospheric Environment (1967)* 16, no. 1 (1982): 2725-34.

Fisher reports measurements of pH and sulfur in rain from a number of sites in Ireland. The report compares the results obtained for Ireland to data for sites in southwestern Norway that are free of industrial sources of sulfur. The report also attempts to model the results and determine the sources of acidity and sulfur.

The models indicate that 50 to 65 percent of all sulfur deposited in Ireland comes from Great Britain. The highest levels are found in the east and decrease toward the west of Ireland. Careful analysis also showed that sulfur was not detectable in the westerly winds coming from the United States. Sulfur depositions from local sources were found to be the highest in areas near airports.

The average annual pH of precipitation was sampled at eight sites in Ireland. It was found that no change in the average annual pH was observed for the years 1968 to 1974. The most acidic precipitation was observed at the site in Clones where the pH values ranged from 4.8 to 5.3, with the most common value of 4.8. The least acidic sample site was Birr, where the values ranged from 5.7 to 6.2, with the most common value of 6.2. A comparison of all ions in the precipitation at these sites in 1972 found that the sulfate concentration at Clones was twice that at Birr (40 ppm versus 20 ppm). Nitrate values were similar at the two sites. The precipitation at Clones had about half the dissolved calcium ions as that measured at Birr. This seems reasonable because calcium ions can counteract the effect of acidic ions such as sulfate. However, rain pH does not tell the entire story because it does not include dry deposition. When dry deposition is added, the average pH for all of the Irish stations over the period of 1968 to 1974 falls to 5.1.

In southwest Norway, ten sampling stations had an average annual acid deposition pH of 4.4. There was no evidence for a significant change in acid deposition in Norway during this period.

The model developed in Fisher's report attempts to determine the relative contributions of wet and dry acid deposition in rural Ireland and compare them to those of rural southwestern Norway. For Ireland, he estimated 1.2 g sulfur/m^2 per year; for Norway the value was about 1.4 g sulfur/m^2 per year. The Irish deposition was about 50 percent wet and 50 percent dry; in Norway, over 66 percent of the deposition was wet. This is much less than the values reported for central England. Table 8 summarizes the data for the three countries.

TABLE 8 Deposition of sulfur in three countries

Country	Deposition Type (g sulfur/m^2 year)		
	Wet	Dry	Total
Rural Ireland	0.6	0.6	1.2
S.W. Norway	1.0	0.4	1.4
Central England	1.0	4.0	5.0

Source: Fisher, 1982, p. 2729.

The modeling of sulfur decomposition suggests that about 50 percent of the total sulfur deposition in Ireland is from human sources. Of that, 65 percent comes from the United Kingdom. This means the United Kingdom is responsible for about one-third of the total sulfur deposition in Ireland that is attributable to human activity. The natural background sulfur has a number of sources including hydrogen sulfide formed in tidal flats and marshes, and sulfate in soil particles and fertilizer.

The same model was applied to the Organisation for Economic Co-operation and Development (OECD) data for southwestern Norway. In this case, only about 20 percent of the sulfur was attributed to natural background; when applied to all of Norway, the background value rose to about 40 percent. Precipitation in Norway is substantially more acidic than in Ireland (pH 4.4 versus 5.1). Part of this difference may be due to the presence of dust and soil particles in the air in Ireland from agricultural practices that neutralize some of the acidity in the precipitation. This could be the source of the observed higher calcium ion concentration in precipitation in Ireland relative to Norway.

Hungary

E. Meszaros, "Les Pluies Acides en Hongrie,"
Pollution Atmospherique 110 (1986): 112–15.
Wet and dry acid deposition was collected in Hungary beginning in 1968. Two types of collectors were used; some were open collectors, and some collected and measured the samples automatically. The network covered all regions of Hungary.

Analysis of samples for the six summer months of 1984 show the spatial distribution of pollution. The lowers pH values occurred near the city of Szombathely on the border with Austria and Slovenia. The average pH at this collector was 4.0. The average pH rose to 4.5 farther east along a line running from the northern border at Győr, south through Keszthely and to the southern border. Farther west, the pH rose to 5.0 along an arc that circles around Budapest from the southern border around to the eastern boarder at Kisvarda.

The region around Budapest had two sampling stations. The open collector recorded an average pH of 5.3, and the automatic collector showed an average of 4.9. Farther east, the average pH was higher, with values between 5.3 and 5.8. However, two automatic collectors had values of 4.0 and 4.6.

The open collector site data for 1968 to 1970 was compared with automated collectors for 1977 to 1980. The electrical conductivity decreased by a factor of 2 over this period, indicating a reduction in total dissolved pollution. Calcium ions decreased from 4.4 to 1.7 ppm, presumably indicative of reduced dust. Nitrate fell from 0.83 to 0.58 ppm, and sulfate fell from 3.7 ppm to 1.9 ppm over this period. All these changes showed a general reduction in pollution during this period. Due to the use of two different collection methods, there was some uncertainty about the interpretation of the meaning of the results.

Air was also analyzed to determine the concentration of various pollutants. Table 9 shows the data for these pollutants, and it is clear that the primary acid source in the air was sulfur oxides and sulfuric acid.

TABLE 9 Concentration of pollutants measured in air samples

Pollutant	Sulfur Dioxide	Nitrogen Oxides	Nitric Acid	Sulfate	Nitrate
Concentration ($\mu g/m^3$)	8.0	2.1	0.82	2.8	0.88

Source: Meszaros, 1986.

TABLE 10 Wet and dry deposition of sulfate and nitrate

Substance	Dry (g/m² per year)	Wet (g/m² per year)
Sulfate	1.1	1.1
Nitrate	0.47	0.33

Source: Meszaros, 1986.

The relative contributions of wet and dry deposition were also determined, as shown in Table 10. Sulfate pollution was evenly divided between wet and dry deposition, and slightly more nitrate came from dry deposition than from wet. The hydrogen ions were not measured in the dry deposition.

Acidification in Hungary has been causing damage to forests and agriculture. Damage to forests is due to both direct damage from the deposited acids and due to the mobilization of toxic inorganic aluminum ions from soils.

Austria

S. Smidt, "Untersuchungen über das Auftreten von Sauren Niederschlägen in Österreich" [Investigations about the Occurrence of Acid Precipitation in Austria], *Mitteilung der Forstlichen Bundesversuchsanstal Wein* 150 (1983): 1–92.

Samples of snow and rain were collected during the winter of 1981–1982 and the summer of 1982 in areas with and without proximity to industrial pollution. Samples were analyzed for pH, electrical conductivity (a measure of total ion concentration), sulfate, nitrate, chloride, magnesium, calcium, and zinc. The pH was measured with and without the effect of dissolved CO_2. The free acid in samples was determined both from the pH and from the total ion balance of samples. The sample sites in Austria are marked on the map in Figure 9.

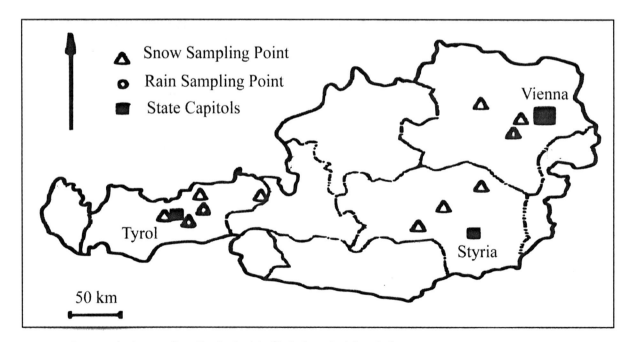

FIGURE 9 Snow and rain sampling sites in Austria. (Data from Smidt, 1983.)

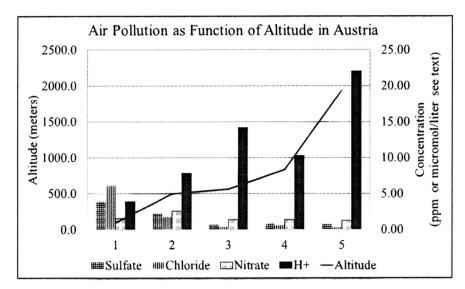

FIGURE 10 Snow data as a function of altitude. Sulfate, nitrate, and chloride are shown in ppm, and hydrogen ion concentration in micromoles/liter. 1=Innsbrück, 2=Vill, 3=Seilbahn Bodenstation, 4=Heiligwasser, and 5=Klimahaus. (Data from Smidt, 1983.)

Most samples, especially snow samples, showed lower than natural pH values. Samples that were close to natural pH had high concentrations of neutralizing ions (calcium and magnesium), indicative of the fact that acids had been neutralized in these samples by dust.

Samples from high elevation, Patscherkofel, Achental (Tirol), and Hochfilzen (Tirol), had distinctly lower pH values and also low negative ion concentrations. The data for snow sample pH as a function of altitude are shown in the Figure 10. Note that the upper bar in each pair for acid concentration increases with altitude, but the negative ion pollutants (lower bar) are greatest at the lowest elevation and decline with elevation. Chloride ion (Cl⁻) has no effect on pH but is included in Figure 10.

Samples from lower Austria, which was thought to be a relatively pollution-free area, had higher levels of pollution than the samples from the areas with high emissions. The data for the area around Vienna are shown in Figures 11 and 12. The data in Figure 11 show snow sample pH values, including the contribution of CO_2 to the acidity. Figure 12 shows the sulfate concentration for the same sites. Note that the samples from Hollabrunn have the highest sulfate but show relatively high pH values. The area west of Vienna and Baden has very low sulfate but also low pH, indicating acid pollution.

The import and export of pollution was not part of this study. The unexpected results for the areas distant from obvious pollution sources indicate that more study was required to fully understand the sources of the observed acidity.

Based on this study, it was concluded that acid pollution is a problem for the whole of Austria and that the decline of the pine and fir forests was a result of acid pollution.

West Germany

W. Knabe, "Anzeichen einer großräumigne Beeinträchtigung der Wälder in Nordrhein-Westfalen durch Einwirkungen von Luftverunreinigungen" [Indications of Large Scale Forest Damage in North Rhine-Westphalia as the Result of the Impact of Air Pollution], *Wissenschaft und Umwelt* 2 (1984): 51–71.

North Rhine-Westphalia is one of the most economically important areas of West Germany and home to a great deal of heavy industry, especially in the Ruhr region with its major steel industry and coal mining.

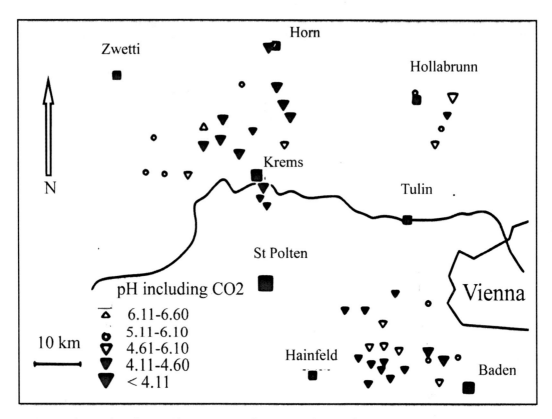

FIGURE 11 The pH data for sampling sites near the Austrian border. (Data from Smidt, 1983.)

There were early indications of the harm being done by air pollution as early as 20 years ago, but the damage did not become obvious until 1982. In the mid-1960s there was a large increase in energy consumption, and damage from ozone was noted in 1967 and 1968 among sensitive bio-indicator species.

This study used lichens and Norway spruce as indicators to evaluate the impact of air pollution. Air pollutants were found to accumulate in needle tissue outside polluted areas, and acid deposition was found even in remote areas. Root damage was observed along with increased concentrations of heavy metals in forests far removed from the sources of pollution. The pollution had a negative impact on the health of Norway spruce, and deciduous trees also suffered bark damage.

The study examined the penetration of acids into various levels of soil and the interaction of wet and dry deposition on trees and soils.

The region was divided into four zones: the overload zone, the load zone, the screening zone, and the outer sink zone. The overload zone is centered on the Ruhrgebeit, the major industrial zone on the Ruhr River. The degree of forest damage was greatest in the overload zone and the least in the outer sink zone.

The study concluded that the only way to reduce the damage to forests is to reduce emissions.

H.-W. Georgii, C. Perseke, and E. Rohboch, "Wet and Dry Deposition of Acidic Components and Heavy Metals in the Federal Republic of Germany," in Commission of the European Communities, *Acid Deposition: Proceedings of*

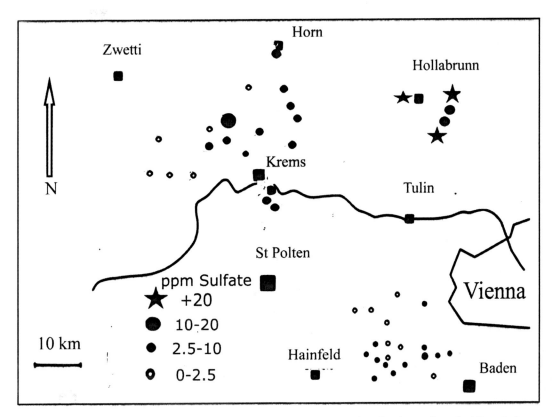

FIGURE 12 Sulfate concentration in sampling sites near the Austrian border. (Data from Smidt, 1983.)

the CEC Workshop Organized as Part of the Concerted Action "Physico-Chemical Behaviour of Atmosphere Pollutants," Held in Berlin, 9 September 1982, ed. S. Beilke and A. J. Elshout, 142–48 (Dordrecht: D.Reidel/Kluwer, 1982). Thirteen sampling stations covering all areas of the Federal Republic of Germany were studied from 1979 to 1981. Wet deposition was collected using plastic funnels and bottles. Dry deposition was collected on glass sample surfaces. The samples were analyzed for pH, sulfate, nitrate, chloride, and heavy metals (lead, cadmium, iron, and manganese).

For rain samples from polluted areas such as the heavily industrialized Frankfort/Main area, 50 percent of all samples had pH values of less than 4.2. In less polluted areas such as Hof/Saale, 50 percent of the samples were less than pH 4.6. The sulfate contribution to the acidity was 55 to 60 percent of the total, with nitrate making up 25 to 30 percent. Chloride was less than 15 percent except at the sampling site on the Baltic Sea.

Wet deposition can be evaluated both in terms of the concentration of the pollutants and the total amount deposited. The highest concentrations are found in light rain where the pollutants in the air are concentrated in only a small amount of rainfall. On the other hand, a heavy rainfall can contribute more total pollutants owing to its longer duration.

The deposition of pollutants was highest in the summer, and the highest rates of deposition were in the industrial Ruhr region, where the rates were as high as 9.2 mg sulfur/m^2 per day in the summer. However, in Duselbach, an area thought to be unpolluted, the rates of deposition were 50 to 60 percent of the values of the Ruhr area. Nitrate pollution followed a similar pattern, with values as high as 3.6 mg nitrogen/m² per day in Hamburg. Values in less polluted areas such as Duselbach and Hof were about half the maximum observed values.

Dry deposition measures only particles and not gases. The dry deposition of sulfate varied from 9 to 29 percent of the wet deposition. Although the gases were not measured, it can be assumed that in the heavily polluted regions the total dry deposition exceeds the wet deposition.

For heavy metals such as lead, wet deposition accounted for the majority of the deposition. The total deposition of lead ranged from 40 µg Pb/m² per day in the unpolluted areas to from 87–100 µg Pb/m² per day in the most polluted Rhine-Main area. The cadmium levels were 1–4 µg cd/m² per day.

The deposition of iron and manganese was primarily in the form of dry deposition, as the relatively large particles fall out of the air. As a result, pollution by iron and manganese was localized near the sources. Because lead and cadmium were primarily deposited in precipitation, their distribution followed the prevailing weather patterns.

P. Winkler, "Trend Development of Precipitation pH in Central Europe," in *Acid Deposition*, 114–21 (1982).

Data were taken from various literature sources covering 50 years of rain samples in the Federal Republic of Germany. Over the entire period, samples close to pH 4.2 were observed. In Hamburg an automatic sampler was used to obtain a multi-year record, which was used to examine seasonal variations and the effects of wind direction on the pH of rain. In addition to pH, the electrical conductivity of samples from Hamburg and three stations in Scandinavia was also measured.

The lowest rain pH that has been reliably reported is around 3.5, but most samples are at pH 4.2 and higher. By analyzing the relationship of pH and conductivity, it was concluded that the gaseous sulfur dioxide dissolved in rain aerosols is limited. Rain with pH values less than 4.2 can form only under very specialized conditions. This is further supported by the observation that rain pH was essentially independent of wind direction and season. Because the emission of sulfur dioxide in winter is roughly twice as great in the Hamburg area as in summer, one would expect about a 0.3 pH decrease for rain in winter over summer. The fact that this change does not occur supports the idea that the formation of acid rain by gaseous pollutants is limited.

The consequence of this is that the ability of rain to purge the atmosphere of sulfur dioxide is limited; when more sulfur dioxide is present in the air than can be dissolved in the available rain, the pollution spreads out and the area affected by the pollution increases. The sulfur dioxide levels in Central Europe had reached levels needed to produce rain of pH 4.2 by 1935. Since then, the total emissions of sulfur dioxide have doubled. This has not caused the rain to become more acidic but rather to spread the pollution to more distant areas. The resulting conclusion is that emissions would need to be reduced well below the 1935 level to raise pH above 4.2, and the primary effect of reducing emissions will be to reduce the distance they travel and the size of the area affected by a given source.

Norway and Great Britain

P. F. Chester, "Acid Rain, Catchment Characteristics and Fishery Status," *Water Science and Technology* 15 (1983): 47–58.

The relationship between inputs of acid from precipitation, water composition, land use, and status of fisheries was studied in the Solandet region of southern Norway.

A recent report has observed that even though the base rock in this area is primarily granite, this type of rock can still neutralize acid from precipitation and has adequate neutralizing capacity to neutralize hundreds of years of precipitation at pH 4.3. The rate of neutralization by granite should be adequate to neutralize the majority of the acid precipitation, but the extent of neutralization will depend on local water flow characteristics.

The study involved 669 medium and small lakes in the region. The majority of water input to these lakes is from precipitation that has interacted with soil and rock and then flows into the lake.

The lakes were arranged in order of increasing acidity, and the percentage of fishless lakes in each category was determined. The lakes were also arranged in order of increasing sulfate, and the percentage of fishless lakes was determined. The results showed a dramatic inconsistency. The percentage fishless lakes increases as acidity increases, but the percentage of fishless lakes decreases as sulfate increases. If the acidity is directly related to acid precipitation, then the increased sulfate should correspond to more fishless lakes. This means that one cannot directly relate fishless lakes with increased input of acid rain.

Further analysis of the data examined the relationship between hydrogen ion concentration and sulfate and found they were not related to each other. If one assumes the primary source of acidity is H_2SO_4 from precipitation and dry deposition, then the only explanation for the observed differences would be differences in neutralization, varying from 17 to 100 percent of the acid input.

A more detailed analysis of the lakes involved looking at all the positive ions associated with the neutralization process by rocks, calcium, magnesium, sodium, potassium, and aluminum and to total acid inputs from both sulfuric and nitric acid. The fact that bicarbonate ions were absent in the water demonstrated that there were no other sources of neutralizing ions. The results again confirmed that the degree to which fish were lost was not related to acidity; for lakes above pH 4.4, the most barren lakes had the lowest sulfate. Loss of fish also did not correlate well with the amount of toxic aluminum in the lakes.

When the lakes were divided into five sets shown in Table 11 based on the percentage fishless lakes, the group with the highest percentage of fishless lakes was the group with the lowest total input of sulfate plus nitrate. There was a clear trend: as the percentage of fishless lakes decreased, the total input of sulfate plus nitrate increased. It is also noteworthy that there was a strong relationship between altitude and the percentage of fishless lakes, with the worst lakes at high altitudes.

There were several limitations to the study. Each lake was sampled only once in the years between 1975 and 1977, but the loss of fish could be due to chemistry from several previous years. The extent of acid due to snowmelt cannot be assessed based

TABLE 11 Chemical properties and fish population in southern Norway

Lake Group	% Fishless	Total Acid Anions $SO_4 + NO_3$ (μeq/L)	Total Hydrogen Ions (μeq/L)	Calcium (μeq/L)	Altitude (meters)
1	67–89	54	24	20	632
2	50–65	71	22	30	508
3	36–55	77	18	38	434
4	24–35	93	12	57	310
5	9	128	4	118	200

Note: The unit μeq/L (microequivalents/liter) is a way of expressing the number of discrete ions per liter related to moles/liter.

on just one sample per year. Also, the growth of sphagnum moss could be changing water chemistry in a negative way that is not reflected in the analysis.

The improved fish stocks in the lower altitude lakes may be due in part to the higher concentrations of the important nutrient calcium. In fact, the authors calculated that if all the rain were pure, it would not dissolve enough rock to supply the calcium needed for healthy fish stocks. Thus, eliminating acidity in rain could have negative consequences.

This study shows the need for additional studies with more samples to determine the true impact of acid precipitation. It also demonstrates clearly that the simple idea that reducing acid rain will improve fisheries is probably not valid.

F. C. Elder, "Acid Precipitation: An Unseen Plague of the Industrial Age" [editorial], *Water Quality Bulletin* 8, no. 2, (1983): 58.

Editorial—Air pollution has been recognized as a problem since people began using fire. The approach to dealing with this problem has included using filters to remove large particles that fall to earth quickly and building taller chimneys to remove the pollutants from urban areas. This suggests that the problems were solved. However, what goes up must come down, and it does that dissolved in rainwater.

The earth has developed a system at equilibrium in which the weak acidity of the rain and surface water is controlled by dissolved carbon dioxide from the atmosphere and weathering of rock and soil by the weak acid in rain. The biological world has arisen and adapted to these conditions. Only a few exceptional places such as bogs and hot springs have developed unique ecological systems.

Combustion gases from human activity include the oxidized forms of carbon, nitrogen, and oxygen. These return to earth by gravity and by dissolving in precipitation. Biological systems utilize the carbon and nitrogen oxides as nutrients, but only small amounts of sulfur oxides are absorbed by the biosphere. As a result, sulfur oxides are the primary cause of excess acidification that can have detrimental or even lethal consequences for biological systems adapted to the preindustrial equilibrium.

The long-range transport of unseen pollutants presents significant challenges. First, the pollution is invisible and in early stages may not produce obvious damage. Second, because the source is so distant, the polluter is difficult to identify and often resists taking expensive action to remove pollutants.

Just as untreated sewage cannot be safely disposed of in rivers, the atmosphere cannot be treated as the place for disposal of acid-forming pollutants. The question for society is not whether we can afford to control pollution but whether the biological systems we depend on can survive if we do not.

I. A. Nicholson and I. S. Patterson, "Aspects of Acid Precipitation in Relation to Vegetation in the United Kingdom," *Water Quality Bulletin* 8, no. 2 (1983): 58–66, 108–9.

The damage caused by acidic air pollution has been recognized for over 100 years, but most recognized damage was thought to occur only close to major sources. In the 1950s plant damage was linked to ozone and other photochemical pollutants. Although plant injury was linked to pollution in areas where a direct connection could be made, there is still no general understanding of the impact of acid rain on plant growth or behavior in field studies. Because the pollutants have historically occurred only at low concentrations and are widely dispersed, the consequences of pollution have remained largely invisible.

The results of the analysis of atmospheric SO_2 are shown in Figure 13. The region of highest atmospheric SO_2 concentrations with values greater than 100 $\mu g/m^3$ are in the industrial

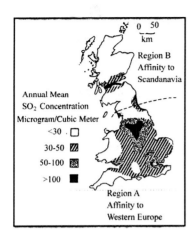

FIGURE 13 Sulfur dioxide concentrations in Great Britain. (Data from Nicholson and Patterson, 1983.)

Midlands of the United Kingdom. A second region of high concentration, 50–100 µg/m³, is found in the Central Valley of Scotland. The pollution from the Midlands is closely connected to a band of highly polluted areas running all the way to Central Europe. The pollution from the Central Valley of Scotland seems more closely linked with Scandinavia. Over 40 percent of the United Kingdom has concentrations of 30–50 µg/m³, and this area includes much of the agricultural land of the United Kingdom.

The areas relatively free of pollution are the Scottish Highlands and Southern Uplands and the Welsh mountains.

Studies in the 1950s linked excess acidity in the English Lake District to air pollution because the hydrogen ion concentration and the sulfate ion concentration increased together.

Studies shown in Figure 14 from 1978 to 1980 found that the weighted average pH of rain across the United Kingdom ranged from pH 4.7 to 4.2. The lowest pH occurred on the southeastern regions with pH gradually increasing as one moved northwest. The acidity was attributed primarily to sulfur oxides (71 percent), with nitrogen oxides making up the rest (29 percent). These values show that acid precipitation in the United Kingdom is comparable to that reported in the Northeastern United States and Scandinavia.

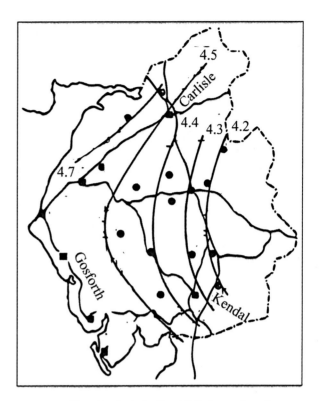

FIGURE 14 The pH of rain in Great Britain, 1978–1980. (Data from Nicholson and Patterson, 1983.)

Deposition of acid pollution on plants can occur in several different ways. Direct contact with polluted rain and exposure to polluted fog are two means of wet deposition. Dry deposition can occur when polluted particles deposit on vegetation. Also acidic gases, SO_2 and NO_2, can be absorbed directly into the stomata (openings in leaves through which plants breathe). Dry deposition on vegetation also dissolves in water that comes in contact with vegetation after dry deposition, increasing the acidity of the water still further.

Studies show that the major route of damage to plants is the direct absorption of the dry gas into the stomata. The sulfur in SO_2 gas may be used to support the sulfur needs of the plants if they are deficient, but in most cases the plants already have sufficient sulfur. Even low concentrations of sulfur dioxide have been shown to impair the growth of plants, even when no visible injury is observed. Nitrogen oxides can also produce injury in this way, and the combination of the two gases together produces a synergistic effect in which the damage is greater than the sum of the two gases alone.

When leaves are exposed to very acidic water (pH less than 3.0), direct, visible injury results. Scotch Pine needles under these conditions show erosion of the waxy surface of the needles. Highly acidic simulated rain water (pH 2.5) caused a reduction in both bud formation and growth in evergreens. When the reduction was compared between plants from historically polluted areas and nonpolluted pristine areas, the plants from seeds from polluted areas showed much less reduction in bud formation than the plants from pristine areas. This may indicate that the plants are adapting to living in polluted areas.

As for the effects on plants and plant ecosystems, in a polluted bog near the city of Sheffield, the smoke pollution may have been a factor in a change in the dominant species. It was observed that Sphagnum (a type of moss) was replaced by a flowering plant called Eriogonum. Another study found that bracken ferns were less fertile when exposed to acidic water (less than pH 4.2).

In addition to the direct interaction of acidic deposition on plant surfaces, there are significant effects of acid deposition on soils, and the absence or presence of plants has a direct impact on the acidity reaching the soil. Studies showed that water reaching the soil after interacting with pine needles was more acidic than the rainfall itself. In one study, rainfall of pH 4.2 became significantly more acidic after interacting with foliage. The water flowing to the ground through needles was pH 3.7. Water coming to the ground within 5 cm of the stem was more acidic still, with a pH of 3.25. Thus, water is picking up hydrogen ions from leaves and stems, possibly from dry deposition on its way to the soil. The water interacting with leaves and stems was also enriched in magnesium, calcium, and sulfate ions. These ions were presumably leached from the plants.

The studies showed that dry deposition had the largest contribution to soil acidification, followed by wet deposition. The third contribution was from leaching from leaves.

The impact of acid pollution on insect and pathogen relationships with plants can be either positive or negative. One study found that acid precipitation less than pH 4.0 greatly reduced the prevalence of black spot on roses. Other studies have found that plants exposed to high concentrations of SO_2 are more attractive to some insects, so the damage caused by the insects will increase.

Conclusions
1. Both wet and dry deposition must be considered to understand the effects of pollution on ecosystems. At present, western European nations have focused more on the effects of the gaseous pollutants SO_2 and ozone on agriculture. In Scandinavia, the focus has been on acid rain and the effects on fisheries.
2. The pollution problem in the United Kingdom can be divided between the north and the south. In the south, there are major sources of pollution and their effects on agriculture are the primary concern. In the northern area, there are few gaseous pollutants and the terrain, geology, and land use are similar to those in Scandinavia.
3. In the northern area of the United Kingdom, there is a significant decrease in the pH of rain as one moves west to east. The total amount of acid deposited in important timber growing regions is high, and studies in the United States suggest this may be a significant problem for timber-growing regions. Furthermore, these areas are remote, and it will be difficult to compensate for acidification by adding basic substances such as lime to try to neutralize the soil.
4. There is no clear evidence that acid rain in the United Kingdom is doing significant harm. The studies summarized in this study

were performed under test conditions, but one must expect that similar problems would occur under natural conditions.

S. Barnabas, "Editorial," *Water Quality Bulletin* 8, no. 3 (1983): 114.

Dr. Barnabas is the manager and principle investigator of the World Health Organization Center on Surface and Groundwater Quality in Ontario Canada.

Editorial—There are two major environmental concerns today: acid precipitation and increasing atmospheric CO_2. Both problems are characterized by the fact that the pollutants are transported over large distances, and have long-lasting effects.

While the majority of scientists report the damage done by acid pollution and call for immediate action to reduce emissions, a minority call for more studies to obtain definitive proof before taking expensive actions to reduce emission. The minority point to decreasing pH in remote places like the Antarctic as a way to strengthen their argument that the sources of acidification may be a natural process rather than human emissions.

The "acid test" that this is a problem caused by human activities is the rapid spread of acid precipitation in the past twenty years. In northeastern regions of the United States, the area effected has tripled during the past twenty years. The same observations are now being made in Scandinavia and northern Europe. There have not been any significant changes in the past twenty years in the natural processes that produce acid rain, such as volcanic activities. So the problem is created by human activities.

R. F. Wright, "Acidification of Freshwaters in Europe," *Water Quality Bulletin* 8, no. 3 (1983): 137–42.

Fish kills of salmon caused by acidic water were first reported in the early 1900s by Norwegian fisheries inspectors. By the 1930s, fish hatcheries had begun using limestone to neutralize acid water. By the 1950, seven major rivers in southern Norway no longer had salmon, and many barren lakes were reported. A publication in 1955 proposed the idea that the acidification of surface water was the result of acid precipitation. A 1968 publication by Olden noted that similar damage was occurring in southern Sweden and stated that the only reasonable explanation for acidification was the increasing emissions of acid-forming pollution, namely sulfur dioxide.

The 1978 report "Long Range Transport of Air Pollutants" by the Organisation for Economic Co-operation and Development (OECD) documented the extent of the problem. It found that precipitation was ten to thirty times more acidic than normal over most of central Europe plus the eastern half of Great Britain and the southern half of Scandinavia. Air pollution by sulfur and nitrogen oxides doubled between 1950 and 1970, and the European monitoring network documented roughly the same increase in acidity of precipitation.

The sensitivity of lakes and streams to acidification depends on the underlying soil and rock. Areas with little carbonate (limestone) are most susceptible. These conditions predominate in Scandinavia, parts of Holland and Denmark, large parts of Germany, and northern Great Britain.

As acid precipitation is added to surface water, the first reaction is the neutralization of the dissolved bicarbonate (HCO_3^-), which normally buffers surface waters. The pH remains above 5.5 as long as there is bicarbonate present. In this process, sulfate replaces bicarbonate as the negative ion associated with positive calcium and magnesium ions. When the buffering ability is exhausted, the pH drops below 5.5. As the addition of acid continues, aluminum becomes soluble, and the pH drops to 5.0 and below. Aluminum is a primary toxin for fish and is especially toxic for hatchlings and juvenile fish.

Acidification has eliminated fish from an area of 13,000 km² in southern Norway shown in

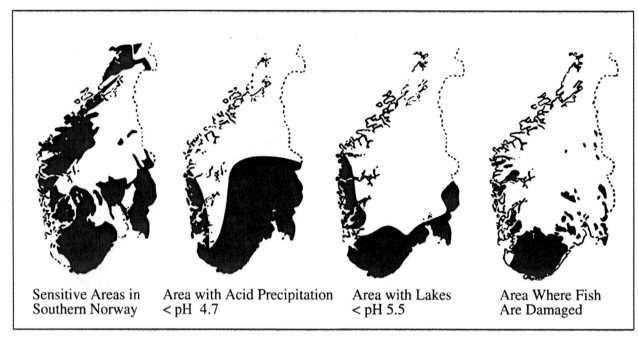

FIGURE 15 Acidification of lakes in Norway. (Adapted from Wright, 1983.)

Figure 15 and dramatically reduced populations in another 20,000 km². The same problem has been observed in western Sweden. Recently, fish kills have been observed in areas of Norway with no previous record of fish loss. In Sweden, of 85,000 lakes larger than 1 hectare, 18,000 are acidified, and 900 have damaged fish populations, as shown in Figure 16.

Finnish lakes and rivers also have problems with acidification. Unfortunately, the Finnish program to monitor water quality lacks the ability to properly measure the extent of the problem. Further research is needed to determine the scope of the problem in Finland.

The extent of acidification in Great Britain has not been well studied, and the scope of sensitive regions is not yet surveyed. However, at least one significant acid-related salmon kill has been reported on the Esk River, which drains the Lake District. Other evidence of acidification due to acid pollution has also been noted in the Lake District. There are indications that many of these lakes were already acidified in the 1950s and that precipitation with pH values of 4.5 and less is common. Problems have also been observed in southern Scotland.

Central Europe has a number of potentially sensitive areas, and there are few data on the extent of acidification there. The potential problem areas include the Vosges mountains of France, the Alps of Switzerland and Austria, the Black Forest (Schwarz Wald) in Germany, and the Tatra mountains of Czechoslovakia.

Acidification of lowland lakes has also been observed in Denmark, the Netherlands, and Belgium, but no detailed analysis has been conducted.

Although acid emissions have remained fairly stable over the past decade, acidification has continued. This suggests that the ecosystems have not yet reached a steady state with the acid emitted in the 1950s and 1960s. If no emission controls are begun, sulfur dioxide emissions will increase by at least 30 percent by 2002. This increase will prob-

FIGURE 16 Acidification of lakes in Sweden. (Adapted from Wright, 1983.)

The first observation was that there was a significant difference in pollution between rain due to weather fronts moving in from the ocean when compared with rain during storms. Storm rain contained more pollution than ocean-derived rain. The concentration of sulfate and nitrate was higher in storm rain (sulfate 6.4 versus 3.4 mg/L), and the total pollution deposited per square meter was higher in storm rain as well (49 versus 8.2 mg/m^2).

A second observation was that acidity increased as the number of days since a rain event increased. This is reasonable because rain cleans the air of pollution; the longer the period since the last rain, the more air pollution will have built up for the next rain event.

The acidity, sulfate, and nitrate in rain was found to increase depending on the distance of the sampling station from the ocean. This is consistent with the observation that ocean rain is less polluted. The general movement of weather is from west to east, and the farther inland one goes, the more pollution sources are present to pollute the rain. For the easternmost sampling site at Phalsbourg, 50 percent of the rain samples had pH values less than 4.5. The Abbeville station located close to the coast had 50 percent of samples with pH values of less than 5.4, indicative of relatively unpolluted rain as the pH of pure rainwater should be 5.6.

Over the period of the study, the relative importance of nitrate seemed to increase. In 1977, sulfur pollution made up 65 to 85 percent of acid pollution. By 1983, the range had declined to 50 to 80 percent sulfur. The trend was the same for all six BAPMoN stations studied.

ably cause the area of acidification to spread into southeastern Europe.

France

M. Zephoris, "Pluies Acides en France" [Acid Rain in France], *Pollution Atmospherique* 103 (1984): 159–65.

Eleven stations from the European Monitoring Environment Program and Background Air Pollution Monitoring Network (BAPMoN) were studied from 1977 to 1983. The sampling program covered all regions of France. The samples were analyzed for positive and negative ions, acidity, bicarbonate (alkalinity), and amount of precipitation.

Italy

F. Pantani, E. Barbolani, S. Del Panta, and F. Bussotti, "Rilevamento di piogge acide in comprensori della Toscana" [A Survey on Acid Rain in Tuscan Sites], *Ressegna Chimica Maggio-Giugno* 35 (1984): 135–41.

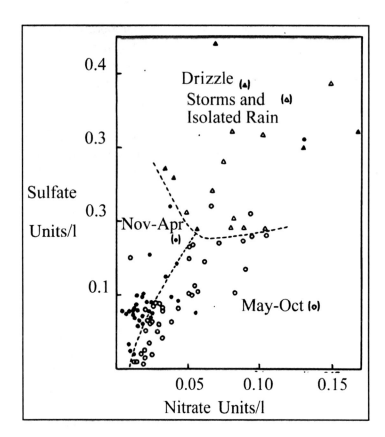

FIGURE 17 Seasonal variation in acid rain in Italy. (Adapted from Pantani et al., 1984.)

The pH of rain was studied at three sites: 1) San Rossore on the west coast, 2) Florence, and 3) Vallombrosa located in the mountains east of Florence. The pH values obtained were primarily in the range of 4.0 to 4.4. During hot summer periods, some neutralization by dust was observed. The primary source of acidity is sulfur dioxide, but the influence of nitrogen oxides increased during the summer in recent years. The chemical composition of rain varies considerably between drizzles, heavy rains, and extended rainy periods. The data for extended periods of rain are shown in Figure 17 in circles, with open circles for summer rain and closed circles for winter rain. The closed triangles are drizzle, and the open triangles are brief rainstorms.

The data in Figure 17 show seasonal variation between summer and winter with the dotted line rising from the bottom. Data for November to April are shown in the open symbols and for May to October in the closed symbols. The variation between slow drizzle events and storms and isolated rain and extended rain events is shown by the V-shaped line in the middle. Extended rain events almost always lie below the V-shaped line at lower pollution concentrations whereas the short-term eventsand drizzle lead to a higher concentration of pollution, as shown by the triangle points above the line.

The horizontal axis is the nitrate contribution to acid, and the vertical axis is sulfate. Note also that the ranges are different, reflecting the fact that there is more sulfate than nitrate under all conditions. The dark circles in the bottom left corner show that nitrate is very low in winter and rises in summer as shown by the open circles. Note that

TABLE 12 Change in runoff ions from 1892 to 1976

Ion	Runoff in 1892 (quantity/m^2 per year)	Runoff in 1976 (quantity/m^2 per year)
Nitrate	5.6	40
Sulfate	36	170
Bicarbonate	360	252

Source: Paces, 1985.

most of the points lie on opposite sides of the dotted line. The summer squalls and drizzle points all lie above the top dotted line and at higher nitrate concentrations than the results for extended periods of rainfall shown by the circles. The pH of rain appears to rarely go below pH 4.0. However, when more pollution is emitted, the area affected becomes larger.

Eastern Europe

T. Paces, "Sources of Acidification in Central Europe Estimated from Elemental Budgets in Small Basins," *Nature* 315 (1985): 31–6. [*Note*: Although the article is dated 1985, the data are from 1982, so this article may be used.]

The acidification of soils in the Elbe River region near Prague in Czechoslovakia was studied. The study concluded that dry deposition of sulfur dioxide and the application of chemical fertilizer were the primary sources of acidification. Direct inputs of acidity in precipitation were of less significance.

In Eastern Europe, the past century has seen a large increase in acidification through air pollution and the use of chemical fertilizers. The results have been decreased growth of forests, corrosion of bridges and buildings, acidification of agricultural land, and the release of heavy metals and aluminum into streams and rivers.

Four separate basins were studied that all had similar bedrock, soil, and climate but differed in sources and amounts of pollution. The data in Table 12 show that the largest increase is for sulfate. The decrease in bicarbonate is indicative of a reduction in the buffer capacity of the runoff, making it more sensitive to acidification.

Air pollution data for the four basins is shown in Table 13. Basin X-14 is a forest facing a major industrial area with open-pit coal mines and power plants burning high-sulfur coal. The other three basins are forests in a rural area without local pollution sources.

These four basins were chosen to allow comparison of three variables: slope of the land, pollution, and agricultural practices. Basins X-0 and X-8 are both rural basins and differ mainly in the slope of the land, with slopes of 3.8 and 13 percent, respectively. Basins X-8 and X-14 have similar slope but

TABLE 13 Total deposition of sulfate and nitrate in four basins

Basin	SO_2		NO_x	
	Deposition ($\mu g/m^3$)	Monthly Variation	Deposition ($\mu g/m^3$)	Monthly Variation
X-0, X-7, and X-8	8.3	37%	11.2	31%
X-14	112	52%	33	38%

Source: Paces, 1985.

TABLE 14 Comparison of streams in four basins

Characteristic	Basin			
	X-0 Forest Low Slope	X-7 98% Agricultural Land Low Slope	X-8 Forest High Slope	X-14 Forest High Slope near Industrial Area
pH				
Rain	4.27	4.27	4.27	4.19
Runoff	6.84	7.14	6.15	4.90
Hydrogen ions				
Rain	0.42	0.37	0.40	0.46
Runoff	0.000	0.000	0.001	0.054
Sulfate				
Rain	12.2	10.5	12.0	19.6
Runoff	8.0	9.0	28.0	96.0
Nitrate				
Rain	4.3	3.7	4.0	5.5
Runoff	0.56	0.58	19.0	12.0
Aluminum				
Rain	0.79	0.68	0.74	1.0
Runoff	0.22	0.11	0.06	3.2

Source: Paces, 1985.
Note: All values other than pH are kilograms/hectare/year.

differ in their exposure to air pollution. Basins X-0 and X-7 differ in that X-7 is primarily agricultural rather than forested, but both are rural and of similar slope.

Rain and snow samples and stream runoff were collected and analyzed. The results are shown in Table 14.

A detailed analysis of the sources and sinks of sulfur in the four basins was done to account for the effects of human use, biological processes, and natural weathering. The results showed that the primary input of sulfur in X-7 was chemical fertilizer and in X-14 was gaseous deposition of sulfur dioxide. Basin X-14 had more sulfate in rain than the other two regions, and the dramatically higher sulfate runoff is due to this dry deposition.

The high nitrate runoff in the agricultural basin is due to fertilizer runoff. The high runoff in the industrial area cannot be explained by fertilizer runoff and must be due to the loss of nitrate from the forests stemming from the acidity of the runoff water.

Aluminum in rain is attributed to the presence of dust, which dissolves in the rain to produce aluminum ions.

The rural and industrial basins are separated by 160 km, and the gradient in pollution corresponds to changes in forests and the weathering of rocks and erosion of soil nutrients. Forests in the industrial region show severe dieback while those in the rural region are generally healthy. Erosion was faster in the higher slope areas, but X-14 suffered over twice the rate of weathering and loss of nutrients compared with X-8, a rural forest of similar slope.

The increased acidity of the runoff in the industrial region reduced the buffer capacity of the surface water to below the detectable level, making these waters susceptible to acidification.

The two rural forested areas were able to absorb much of the acid and sulfate from the precipitation as shown by the lower amounts in the runoff compared with the rain. The agricultural and industrial areas were not able to do this.

SOPHIA CONFERENCE NEWS

V. Rich, "Immediate Action Imperative," *Nature* 308 (1985): 501.

Poland is facing a possible ecological catastrophe, according to Dr. Antoni Kuklinski, head of the Committee of Spatial Development of the Polish Academy of Sciences. Dr. Antoni identified three areas that had reached the "breaking point": (1) the natural environment; (2) the technical infrastructure, which includes transport, power supplies, and water management; and (3) the poor conditions in large urban areas.

If industrial investment were cut to 30 percent of the national total this situation could be improved. At the same time, the housing sector must learn to make better use of the available resources. According to Dr. Kuklinski, writing in the government newspaper *Rzeczpospolita,* the whole structure of investment must be changed, with greater powers given to local "self-government" bodies. Investment in the environment must be of primary concern.

This is not the first concern about the environment to emerge in the Polish media. Recently, Warsaw Radio noted that in the upper Silesian industrial belt, 430 out of 100,000 people died prematurely because of environmental conditions and this death rate of 50 percent higher than the national average. Circulatory diseases, cancer, pulmonary diseases and infant mortality occurred much more frequently in this area of Poland. One doctor recommended that people not work in this area for more than 20 years. Those not working should live elsewhere.

This view is reinforced by a report which includes the findings of a seminar held in Lublin in the past fall on the chemical threat to the Polish environment. According to a survey carried out in 1982 by the Planning Committee of the Council of Ministers, this area encompassing a population of 11 million (30 percent of the total population) is under severe ecological threat.

Special attention is focused on the classification of air pollution in terms of sulfur dioxide toxicity equivalent, since recent studies have made Polish planners aware of the problem of sulfur dioxide emissions. The hazard from heavy metal dust from the metallurgical and engineering industry is equivalent to 13.8 million tons of sulfur dioxide per year. Gases from the chemical industry have a sulfur dioxide equivalent of 1.26 million tons. The annual emission of sulfur dioxide by Polish industry is 2.45 million tons.

The report stresses that the whole pollution problem was studied afresh in 1982 and that actions taken to reform the economy and focus on the three strategic problems of food supplies, housing and restoration of the ecology will be introduced during plans for 1986-1990.

L. W. Blank, "A New Type of Forest Decline in Germany," *Nature* 314 (1985): 311.

The forests of West Germany are suffering from rapid, widespread decline. It is estimated that at least half of all forests are damaged to some extent. The damage does not appear to be directly related to acid rain. Rather, the decline may be due to a combination of ozone, acid mist, and changes in climate.

Forest damage observed in the 1970s was limited to small areas and a few species of trees. The current damage is widespread, especially in the Black Forest and Bavaria and includes more species of trees. Similar damage has now been observed in Italy, Poland, Czechoslovakia, and Switzerland. Damage to Silver Fir began in 1976 and this is the most damaged species. Norway spruce is also damaged, and this tree makes up 40 percent of German forests. Damage to monitored spruce forests in Baden-Wurttemberg increased from 6 percent to 94 percent in the two years from 1981 to 1983. Initial damage was most evident at high elevations, but has spread. Older trees are generally the first to suffer, but the patterns of damage are not consistent.

Forest damage inventories were carried out by each state in 1982, 1983, and 1984. There was little increase of severely damaged and dead trees across these studies, but there was a major increase in the areas showing slight damage and moderate damage. The former went from 25 percent to 33 percent and the latter went from 9 percent to 16 percent in one year from 1983 to 1984. There are some uncertainties in these results because it is not clear how much needle loss is normal on a year-to-year basis and needle loss was one of indicators used to measure forest damage.

The initial cause of the decline was attributed just to "acid rain." More recently, scientists have begun to realize that multiple forms of air pollution in addition to just sulfur dioxide and acidity are involved. The fact that there has been no recent decrease in rain pH or increase in sulfuric acid and nitric acid in the rain in the forest areas suggests that acidity alone is not the explanation.

Soil acidification and loss of the nutrient ions calcium and magnesium have also been proposed as the source of the problem. However, the fact that the problem exists in regions with high-calcium soils as well as low-calcium soils suggests this is not an adequate explanation.

Direct exposure to gaseous sulfur dioxide and nitrogen oxides has also been ruled out as the cause. Although these have caused damage in Czechoslovakia, most areas of German forests are not exposed to these pollutants at high enough concentrations to damage the trees. A second indication that gaseous pollutants are not the problem is the fact that the damaged areas have large growth of lichens, and lichens are known to be very susceptible to damage by these gases.

The most widely accepted explanation at present is that ozone due to increasing emissions of nitrogen oxides reacting with the hydrocarbons naturally produced by evergreen forests is causing the damage. Ozone concentrations in southern sections of the Black Forest have been measured as high as 100 μg/m^3. Few monitoring stations exist for ozone, so few data are available. One station in the Baltic Sea, which has not been affected by local pollution, showed a 60 percent increase in ozone from 1956 to 1977. Two stations in forests in West Germany also showed the same general trend. Ozone concentrations were also found to be related to altitude, with higher ozone concentrations at higher altitudes. This correlates with the observation that tree damage was greatest at higher altitudes.

The 1970s were characterized by several hot dry summers that may have stressed trees and made them more susceptible to both air pollution and fungal infections.

The sudden forest decline has prompted a number of research projects. Over the next few years, these should determine whether the ozone hypothesis is correct or whether there are different or additional factors.

N. von Beeman, "Acidification and Decline of Central European Forests," *Nature* 315 (1985): 16.

Tomas Paces in *Nature* reports one of the most complete analyses of the acidification of forests in Czechoslovakia. The area of study was a forest that had lost all of its spruce trees. This region is thought to be similar to other highly industrial areas in Eastern Europe.

Dry deposition of sulfur dioxide was responsible for 75 percent of the acidification and wet deposition for 15 percent in industrial areas. These areas have three times the acid level of remote forests; however, drainage water from the industrial areas is partly neutralized by the effects of the acid dissolving rocks (weathering).

There is general agreement that acidification of surface waters kills fish populations, but the impact on forests is more controversial. Direct damage by pollutants on vegetation is significant, but the larger problem for forests may be the loss of the nutrient ions calcium and magnesium from soil and the release of toxic aluminum. The acidification of agricultural land can be neutralized with lime, but liming forests is not practical and may cause other problems for trees.

E. Collins, "Upwind vs Downwind," *Nature* 317 (1985): 377.

The environmental group Earthscan released a report that argues that the problem of acid precipitation is spreading to Third World countries. They also argue that buildings and crops are damaged in addition to the forests and fisheries that have been the focus of much of the debate between major polluters and those receiving the pollution.

Downwind countries such as Sweden and Norway are considering action against polluters in the International Court and are discussing offering loans to polluters to install scrubbers.

Earthscan notes that both Brazil and the People's Republic of China have serious problems with acid precipitation due to inefficient combustion and the use of high-sulfur coal.

The problem of acidification of soil goes beyond sulfur dioxide emissions and includes nitrogen oxides from cars and ozone. Also, use of fertilizers in farming can cause acidification.

The research on soil acidification and the effects on forests and fish have not produced a clear consensus on the cause and effect relationships involved. This has allowed policy makers in the polluting countries to avoid taking responsibility.

P. Gambles, "Acid Rain Storm from Norway," *Nature* 318 (1985): 503.

Relations between Great Britain and Norway were strained by the release of a film on acid rain by the Central Electricity Generating Board (CEGB) of the United Kingdom. The Norwegian government responded that the video was based on outdated science. The film suggests that there is scientific uncertainty about whether fish kills in Europe are a result of acid rain. The Norwegian government and some scientists interviewed for the video protest that the interviews were distorted in the editing process.

Norwegian scientists recently presented evidence that 15 to 30 percent of sulfur dioxide pollution in southern Norway comes from the United Kingdom. They argue that a comprehensive approach to stopping what they call a "chemical war" is needed, and they acknowledge that action by the United Kingdom alone is not enough.

Norwegian authorities were disturbed by the way the CEGB dismissed the serious problems caused by acid rain on the basis that the scientific community has not achieved 100 percent certainty on the issue.

R. Walgate, "UK Denies Responsibility for Scandinavian Acid Rain," *Nature* 323 (1986): 191.

Lord Marshall, head of Britain's Central Electricity Generating Board (CEGB), announced that scrubbers would be added to only three British power plants to achieve a 10 percent reduction in emissions by 2000. This decision was based on the results of recent joint studies by British and Norwegian scientists. Marshall claims these studies show that British emissions make up a smaller component of the problem in Scandinavia and that the improvements that would result from reductions in emissions are smaller than previously thought.

Norwegian government officials were outraged by this decision, calling it a slap in the face, especially since the announcement was made while Prime Minister Thatcher was visiting Norway. The U.K. plan calls for reductions of 90 percent in emissions only by 2020, and the Norwegians believe this will be much too late. Scandinavian offices report that 33,000 km² in southern Norway are affected along with 2,500 to 6,000 lakes in southern Sweden.

Lord Marshall focused on the estimate that the British contribution to acid rain in Norway has been reduced from 30 percent to 8.5 percent now that better models and more accurate measurements of dry deposition are available. The British note that at this level Britain contributes less acid than Norway's own small emissions. Norwegian scientists counter that while the overall British contribution is only 8.5 percent, in the most affected regions of southern Norway, it is 16 percent and that it is more than five times the Norwegian contribution in those areas. Furthermore, the reduction in acid deposition in Norway is due to the fact that sulfur dioxide is not converted to sulfuric acid as quickly as thought. This means that the plume of pollution from Britain travels farther than previously thought.

The weakness in the British argument lies in the fact that 40 percent of the total deposition in Norway cannot yet be attributed to a source. As much as 10 percent of this could come from as far away as the United States, but Norwegian scientists believe as much as half is from Britain. These pollutants are in air masses that have been over the Atlantic Ocean too long to allow the exact source to be determined. Some of this may be air recirculated from Britain and other European countries.

Chester, the chief environmental scientist for the CEGB, also reported that there is a large amount of sulfur in various forms in the soil of southern Norway. He has proposed that this sulfur is released from the soil when forests are cut. The cutting of forests releases nitric acid, and the nitric acid converts sulfur in the soil to sulfate. Based on this, Chester opposes any effort to reduce British emissions.

A. Williams, "Control of Nitrogen Oxides," *Nature* 324 (1986): 612.

Governments in the United States, Japan, and western Europe have increasingly recognized nitrogen oxides as a significant pollutant both on their own, in acid rain, and in the production of photochemical smog and ozone.

Four technologies to address nitrogen oxide are under consideration:

1. Using precious metal catalysts. This requires fairly clean exhaust gases free of things that can poison the catalyst. It also requires the careful control of excess oxygen. This method is really suitable only for engines and natural-gas power plants.
2. Adding ammonia to the exhaust gases of power plants. This requires careful control of the amount of ammonia added and the exhaust temperatures.
3. Using ammonia with a catalyst on power plant emissions.
4. Adding methane to "burn" the nitrogen oxides. This requires careful control of the added methane and the temperature.

All these methods produce heat, which must be controlled; the temperatures at which the reactions occur must be considered. New technologies are being developed that promise to meet even the most stringent limits and that can be applied to dirty fuels such as coal.

"Polish Government Getting the Message on Environment," *Nature* 326 (1987): 819.

The Polish government recently moved to prevent radical environmental groups that have formed clubs and committees from bringing too much pressure on the government to reduce Poland's 4 million ton annual emissions of sulfur dioxide from coal plants. The environmental social movement will be organized under the Patriotic Committee for National Rebirth (PRON), which will provide oversight for these groups. PRON was established by the government under martial law in an effort to control the initiatives launched by the Solidarity labor movement.

Two active groups, Freedom and Peace and the Krakow Ecological Club, have been pressuring the government on a number of environmental issues. They will probably not be allowed to join the environmental social movement, but the new government-sponsored group may be able to provide better coordination among groups working for a cleaner environment in Poland.

K. Johnson, "Acid Rain Deposition," *Nature* 328 (1987): 465.

The U.K. Review Group on Acid Rain reported that levels of acid rain declined in the first half of the 1980s in proportion to the decline in emissions. However, areas in the Scottish Highlands and Cumbria continue to have deposition rates comparable to those in southern Norway.

The group also noted that, in spite of the addition of two more monitoring networks, good data on acid depositions are still not available.

"Acidification under Attack," *Nature* 330 (1987): 338.

Politicians from pollution-importing countries such as Norway, which has suffered extensive loss of salmon and brown trout, continue to argue with pollution exporters like Great Britain over whether the cause of this damage is acid rain or the nature of the soils in areas of Norway. The controversy has prompted extensive research that is now making it clear that acid precipitation is responsible for the release of acidic compounds from soils that damage streams and lakes.

Two studies in Norway that are part of the RAIN (Reversing Acidification in Norway) project are producing interesting results. One study in Sogndal is adding acid to a previously unpolluted stream to study the effects of acidification on a pristine ecosystem. The second study in Rishdalsheia has covered a stream with a roof to collect the acid rain falling on a highly acidified stream. The collected rainwater is then purified and added to the stream. The results of these studies indicate that the damage of acid precipitation is, for the most part, reversible if the input of acidic pollutants is removed.

Extensive research on acid deposition of both sulfate and nitrate is under way, some of it part of a joint Norway–United Kingdom research effort to study the effects of acidification. Another major study by the UN Evaluation of Long Range Transport of Air Pollution in Europe (EMEP) program is developing models of emissions and deposition of both pollutants. The models show reasonable agreement between predicted and observed acid deposition based on data supplied each year by the various European governments. Nitrate pollution does not travel as far as sulfate, and the nitrate models are not yet as accurate as the sulfate ones.

Considerable controversy still exists within the Scandinavian countries over the relationship of observed forest damage and air pollution. Norwegian forestry researchers are divided on whether the damage is due to acid precipitation or to changes in climate. Norway is highly dependent on wood, and this is an important question for them. In Sweden, forestry officials are united in their belief that acid rain is responsible for the damage to Swedish forests.

SOPHIA CONFERENCE: TECHNICAL BACKGROUND

The Science of Ozone and Smog

The primary scientific issues in the Sophia Conference are the formation of ozone and photochemical smog from NO_x pollution and the effects of lead pollution on the environment. These are related by the fact that the most effective way to remove NO_x from car exhausts is the use of catalytic converters. Catalytic converters require the removal of lead from gasoline.

Figure 18 summarizes the complicated process of forming ozone. The inputs into this reaction are volatile organic compounds (VOCs) and the presence of any form of nitrogen oxide (NO_x). Light

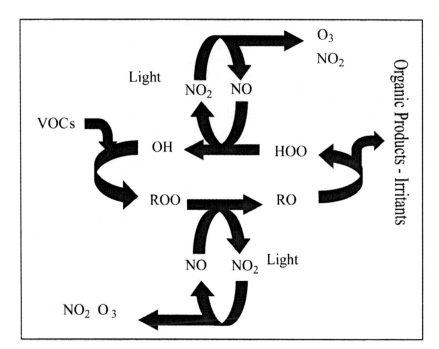

FIGURE 18 Diagram of the processes involved in the formation of ozone and smog.

is required for these reactions to occur. The organic products are chemicals like formaldehyde that are irritants and highly reactive.

Early efforts in the United States to control photochemical smog were focused on removing the VOC part of the chemical equation. Because both VOCs and NO_x are required, it was believed that removing the VOCs would stop the cycle and prevent ozone formation. This would not do anything to deal with the acid rain produced, but most acidity in rain was due to SO_2, so this was considered an acceptable approach.

The United States instituted a wide range of actions to reduce VOCs. Evaporative loss equipment was required in cars to capture gasoline that evaporated from the gas tank in hot weather. Similar equipment was installed at gas stations and gasoline storage facilities. Most oil-based paints were eliminated and replaced with low VOC latex paint. Industries that had once exhausted solvents into the air were required to capture and recycle them. The effort to control VOCs was successful, but the efforts failed to reduce smog as expected.

The reason that VOC control failed to reduce ozone is related to a central concept of chemistry, the *limiting reagent*. The best way to introduce this is to recognize that every chemical reaction is a recipe. So let us consider the recipe for s'mores.

1 piece chocolate bar + 1 toasted marshmallow + 2 pieces graham cracker → 1 s'more

As you have probably made s'mores at some point, you experienced the limiting reagent problem firsthand when you ran out of one of the ingredients. So let us look at how a chemist would analyze the problem.

Each chocolate bar provides 3 pieces of chocolate. Each whole graham cracker has 2 pieces. We go to the store and buy 4 chocolate bars, a bag of marshmallows, and a box of graham crackers. We count the marshmallows and find we have 30 of them. The box of graham crackers has 40 whole crackers. The question is how many s'mores can we make?

You intuitively know that with 4 chocolate bars and 3 pieces per bar, we can make at most 12 s'mores (1 piece/s'mores × 3 pieces/bar × 4 bars × = 12 s'mores), which assumes you can keep people from snacking on the chocolate. (Keep your hands off that chocolate!) We have enough marshmallows for 30 s'mores and enough crackers for 40 s'mores. It always seems to work out this way. You make all the s'mores you can, toast the rest of the marshmallows, and end up with a surplus of crackers. Chocolate bars always seem to be the limiting reagent.

In the recipe for ozone, we saw we needed three things: NO_x, VOC, and sunlight. We cannot control the sunlight, but the United States has tried to stop the reaction by controlling the VOCs. The reason this strategy did not work was that the VOCs were the graham crackers in this process. There are large natural sources of VOCs. Even if people stopped all their emissions of VOCs, there would still be enough VOCs left to use up all the NO_x. The only way to reduce the total production of ozone is to cut out the NO_x.

Health Effects of Ozone

Ozone in the air is linked to a number of health problems, including asthma and heart disease. It is somewhat complicated to report the amount of ozone in the air because the concentration varies during the day.

Figure 19 shows data from three measuring stations. Before 8:00 A.M. there is very little ozone because it has been destroyed overnight in making nitric acid. The concentration peaks in midafternoon and then declines as the sun sets.

The varying concentrations make it difficult to demonstrate a link between ozone and diseases. However, several standard measures of exposure have been defined. The maximum 1-hour average takes the highest 1-hour value and relates that to health effects. One can also do an 8-hour average concentration, which clearly leads to a lower value than the 1-hour value.

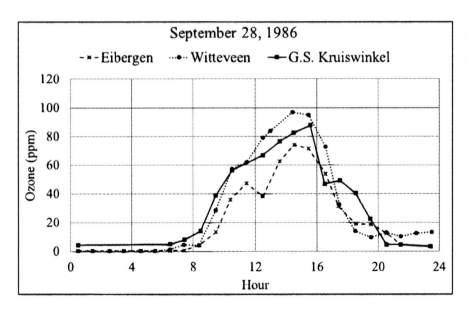

FIGURE 19 Ozone data for three monitoring stations. (Data from Feister et al., 1990, p. 19.)

A report from the World Health Organization found the following health consequences of ozone pollution in Europe.

- Short-term lung obstructions occur after exposure to levels above 160 µg/m^3 for 6.6 hours.
- People with asthma, allergies, or malnutrition are more susceptible.
- Ozone pollution is a risk factor for the development of cardiovascular disease.
- Increased risk of death, especially for the elderly, has been observed at ozone levels as low as 60 µg/m^3 1-hour average.
- Mortality is greater than 20,000 deaths per year.
- Among people over 65 there are 14,000 hospital admissions per year.
- Restricted activity days for people from 15 to 64 years old uses up 60 million person days.
- Respiratory medicine use by children uses up 20 million person days.
- Respiratory medicine use by adults uses up 8 million person days.
- Cough and lower respiratory symptoms in children uses up greater than 100 million person days.[1]

The data in Figure 20 show the rather complicated relationship between ozone concentration in the air and total mortality. At low concentrations, there is a slow rise up to a threshold around 60 micrograms/m^3. Then there is a very rapid rise. The physical explanation for this effect is that the body produces a number of antioxidants capable of removing ozone and preventing damage. It is only when these defenses are overwhelmed that mortality increases rapidly. Also, it is important to

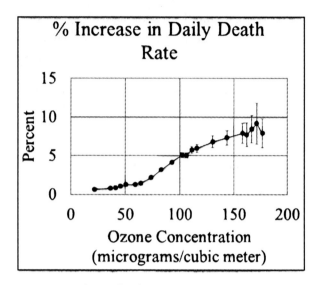

FIGURE 20 Relationship between ozone concentration and daily death rates in twenty-three European cities. (Data from Gryparis et al., 2004.)

note that the level of protection varies with the overall nutrition and health of the individual. Malnourished, elderly, and sick people are affected at lower concentrations than healthy people.

What can be done about acid rain and smog? The solution to the problem of acid precipitation seems obvious: decrease the amount of both sulfur dioxide and nitrogen oxide pollution. The Sophia Conference presumes that the European community has already adopted a treaty on long-range transport pollutants (LRTRP). The focus of these negotiations before Sophia was on steps to reduce sulfur dioxide pollution. No matter how well the previous conferences have succeeded, the treaty does not yet totally fix the problem. The mechanism is in place for LRTRP to continue the process of tightening restrictions as needed. The Sophia Conference is the first opportunity to address the other major issue of acid precipitation from nitrogen oxide pollution. No one advocates doing nothing, but there are several choices to be made, and different countries favor different approaches. There are problems with each solution, and there are ancillary benefits to some possible solutions.

---

1. Adapted from M. Amann, D. Derwent, B. Forsberg, O. Hänninen, F. Hurley, M. Krzyzanowski, et al., *Health Risks of Ozone from Long-Range Transboundary Air Pollution* (Copenhagen: World Health Organization Regional Office for Europe, 2008).

1. Reduce NO_x emissions from automobiles by requiring catalytic converters on new cars.

    Deploying catalytic converters requires a large-scale shift from leaded gasoline to unleaded gasoline. Unleaded gasoline will be more expensive but will also reduce harmful lead pollution. Topics for your research include the expected cost increases for cars and gasoline, and the potential benefits of reducing lead pollution and reducing repair costs for cars.

2. Use lean-burn engines in new cars.

    Lean-burn engine allow the continued use of leaded gasoline. What are the relative costs of this approach and will it work on all types of cars?

3. Reduce NO_x emissions from power plants using new burners, scrubbers, and fuel additives.

4. Use more natural gas to make electricity because it is more efficient and produces little pollution.

5. Build more nuclear power plants, which produce no air pollution.

6. Use less total energy so less fuel is burned.

Reduction of Nitrogen Oxide Emissions

Nitrogen oxide emissions are the most intractable part of the acid rain problem and present one of the biggest environmental challenges, both to technology and to legislation. The majority of sulfur pollution comes from stationary combustion sources such as electric generation plants, which are referred to as *point sources*. The majority of nitrogen pollution, however, comes from the transportation sector and represents millions of *moving sources* instead of a few thousand fixed sources. This makes the implementation of control strategies more complex.

The United States began to implement ternary catalytic converters in the early 1980s. By the time of the Sophia Conference in 1987, all new cars in the United States, including imports, were required to have them. Also, the phaseout of leaded gasoline, which began in the United States in 1978, was completed in 1987. The experience of the United States with catalytic converters and unleaded gasoline provides a wealth of experience you can research as you develop your positions.

West Germany is the only European country that has followed the lead of the United States and passed laws requiring catalytic converters. European automakers with significant exports to the United States have been forced to develop the technology for catalytic converters. This applies primarily to West Germany, which has large export trade with the United States. French and British automakers have fewer exports to the United States.

European automakers, including those in Germany, have applied political pressure to prevent requirements for catalytic converters. They support the lean-burn engine as an alternative to catalytic converters, at least for smaller engines (less than 2 liters). Lean-burn engines use improved carburetors that allow lean fuel/air mixtures. These may also be able to reduce auto emissions of both hydrocarbons and NO_x.

There is considerable controversy over the best technology to solve this problem. The actual costs of implementing the two different approaches are in dispute. Advocates of lean-burn engines say they will cost less than $50 per car and that catalytic converters could add $1,000 or more. Advocates of catalytic converters say the lean-burn engines do not work on large cars and that the costs are greater than $50. They also say that widespread use of catalytic converters, as in the United States, will reduce costs to a few hundred dollars. The truth probably lies somewhere in between.

It is also possible to reduce NO_x emissions from power plants. This is an area where the technology is not as well developed at the start of the game as for SO_2. To some extent, without requirements for pollution control, there is no impetus for technological development. Emissions of NO_x from power plants depend on a large number of factors such as the type of fuel, the burner design, the ratio of air and fuel used, and the overall temperature and design of the combustion area in the boiler.

At the time of this game, the primary way of reducing NO_x pollution from power plants is to replace standard burners with low-NO_x burners. These have been developed and widely adopted in Japan and are beginning to appear in the United States. Europe has not yet begun to use them. There is little doubt that additional technological advancements will be made should there be a widespread demand. The cost of switching to low-NO_x burners is expected to be fairly low. It may be possible to make the changes as part of regular burner replacement. Therefore, European nations are unlikely to resist low NO_x burners for financial reasons as long as they have a reasonable length of time to implement the change.

A summary of the NO_x pollution data for the game is provided in Table 15.

TABLE 15 NO_x pollution for 1985–1986

Country	1,000 Metric Tonnes per Year[a]	Mobile Sources[b] (% of total)	NO_x to GDP Ratio (kg per unit GDP)	Grams of NO_x per Kilometer Driven by Cars[c]
Austria	209	49	1.27	3.76
Belgium	338	55	0.84	2.73
Czechoslovakia	816	—	—	—
Denmark	308	26	1.04	2.48
East Germany	955[d]	—	—	—
Finland	288	53	1.80	7.62
France	1,789	51	1.29	2.81
Hungary	300	—	—	—
Ireland	113	35	—	1.08
Italy	1,917	46	1.06	2.03
Netherlands	599		1.53	3.27
Norway	203	84	2.27	9.65
Poland	1,500	—	—	—
Romania	508	—	—	—
Spain	1,175	46	1.49	4.50
Sweden	437	68	1.48	3.26
U.K.	2,791	45	1.65	2.96
West Germany	2,950	40	1.54	3.73

[a] *Source*: European Environment Agency, 2014, Annex B.
[b] *Source*: These data have been recalculated from EMEP program data: EC-UNEC 2000, www.ciep.at, accessed April 5, 2000.
[c] *Note*: These data do not refer to actual automotive emissions; rather, they are a ratio of total kilometers driven to total national emissions.
[d] *Source*: Iversen et al., 1989, for data in italics.

The Internal Combustion Engine

The majority of cars have used the internal combustion engine since early days of the automotive industry. The basic principle of the engine is that a mixture of air and fuel, usually gasoline, is admitted into a cylinder. Step 1 in Figure 21 shows the mixture being drawn into the cylinder from the left through a valve by the downward motion of the piston. A piston in the cylinder then compresses the air-fuel mixture in step 2. As the piston reaches the point of greatest compression, a spark is generated in the cylinder to ignite the mixture. The resulting explosion drives the piston down in the cylinder, and the energy is captured by the crankshaft to turn the wheels of the car through a series of gears in step 3. Finally the piston moves back up to expel the combustion products through the valve on the right in step 4.

The diesel engine used in trucks and some cars is very similar to the internal combustion engine with one major difference. The diesel engine does not have a spark plug to ignite the fuel/air mixture. Instead, the amount of compression generated by the piston heats the gases to a high enough temperature that the mixture explodes spontaneously without needing a spark. Compressing gases raises their temperature.

Most cars sold in the 1970s and 1980s used a very simple method of generating the fuel/air mixture. The carburetor had a spray orifice in the stream of the incoming air pulled in by the piston. The gasoline was simply sprayed into the stream of air and vaporized by the heat in the intake area on its way to the cylinder. By varying the amount of air and fuel admitted, the engine can go faster or slower. More air and fuel produces faster speed and more power. Most modern cars now use a process called fuel injection. With fuel injection, the air is pulled into the cylinder by the piston and the fuel is injected into the cylinder under high pressure by a pump. Fuel injection gives more control over the relative amounts of air and fuel in the engine, but it adds to the complexity and cost of the engine. A final variation called turbocharging uses a pump to force the air into the cylinders so more air can be available than can be drawn in by the motion of the piston alone. Turbocharged engines provide more power because a larger amount of explosive fuel/air mixture can be placed in the cylinder for each cycle.

For the engine to run, the fuel/air mixture must be explosive. Fortunately, not all combinations of gasoline and air will burn. The range of concentrations at which gasoline will burn varies with temperature and pressure. Under normal conditions, the range for 100-octane gasoline is from 1.7 percent to 7.6 percent fuel vapor in air by volume. If the concentration of fuel drops below 1.7 percent, the mixture is said to be too lean for combustion. If the mixture rises above 7.6 percent, the mixture is too rich for combustion. Within the combustible range, gasoline will ignite if 1) the temperature rises above about 250°C, or 2) a spark occurs.

The burning characteristics of the gasoline in the cylinder depend on how fast the gasoline burns as well as the fuel/air ratio. Gasoline that burns slower provides a smoother expansion of the cylinder than regular gasoline, which explodes very fast. Also, if the gasoline burns slower, it is possible to ignite the mixture earlier in the cycle and get more power from the explosion.

The timing of the spark plug firing is a variable that can be adjusted for the engine. In the 1980s most cars allowed the mechanic to adjust this

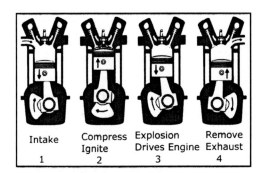

FIGURE 21 Diagram of a four-cycle internal combustion engine.

when the car went in for a tune-up. Modern cars often have this under computer control. The properly tuned engine would ignite the spark at the first possible instant that allows the explosion to keep the crank shaft turning in the correct direction. In practice, the combustion might be slow enough that ignition could begin even before the piston reached the top, referred to in tuning manuals as the "top dead center." (The tune-up manual for the authors' 1970 MGB recommend a timing of 5 degrees before top dead center for optimal performance with 95-octane gasoline.) If the spark plug fires too soon, the fuel ignites and tries to push the piston back down the opposite way. This produces an audible explosion that is often referred to a pinging or knocking. (It sounds like a loud clicking in the engine.) In the extreme, this can damage the engine, and at the very least it can seriously reduce the engine's power.

The timing depends critically on the burning rate of the gasoline as indicated by the octane number. You will see these numbers posted at every gas pump. The higher the number, the *slower* the gasoline burns. These numbers were historically determined by running a small engine on the sample of gasoline and comparing the timing position to standard gasoline samples. The timing angle at which preignition occurs was measured for several standard gasoline samples. The timing angle for pure iso-octane was defined as 100-octane, and the angle for *n*-octane was defined as 0-octane. A graph of the timing angles versus percentage of iso-octane was developed for an engine, then the timing angle for an unknown gasoline was compared to the graph. It is possible to have gasoline above 100-octane if it burns even more slowly than pure iso-octane. Some aviation fuel has an octane number of 140.

In the early days of the automotive industry, the gasoline available was composed mostly of straight-chain hydrocarbons like *n*-octane and thus had a very low octane rating. Because low-octane gas burns so fast, it places constraints on how much it can be compressed in the cylinder before it explodes. It also means that the spark plug timing must be set very late in the cycle. The combination of starting the combustion later and compressing the gas less means that the engine does not run very efficiently. Early cars had serious problems with preignition (knocking). Solutions to this problem are discussed in the section on leaded gasoline.

Combustion Products and Engine Parameters

The nature of the combustion products produced in the engine cylinder change as the fuel/air ratio changes. A rich fuel mixture has more fuel than can burn completely with the available air. This will always result in unburned hydrocarbons in the exhaust. This wastes fuel and reduces efficiency. Alternatively, a lean fuel mixture has more air than fuel and would be expected to produce fewer emissions. However, the fuel in a lean mixture may also not burn completely. When the mixture is too lean, the fuel is so diluted by the air that not all the fuel molecules are ignited by the combustion process.

The data in Figure 22 show the complicated choices faced in optimizing one specific internal combustion engine. The exact values of fuel/air mixture for different engines depend on the compression ratio, but the general trends are true for all engines. The possible range of fuel/air mixtures for gasoline in this engine range from about 10 to 22 percent air-to-fuel by volume. The rich mixture between 11.5 and 13 percent air gives the best power because there is more fuel in the cylinder to burn. However, this range also produces the highest emissions of hydrocarbons that contribute to smog and carbon monoxide (CO), which is toxic. As the air is increased, the engine becomes more efficient in terms of miles per gallon. The best fuel economy occurs at around 15.5 percent air. Note that this mixture also produces the highest amount of NO_x pollution, though the least CO and hydrocarbons. When the mixture

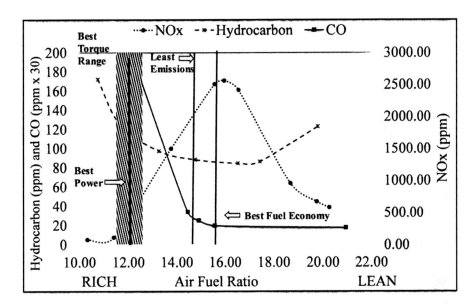

FIGURE 22 Tuning a carburetor for optimum emissions and performance. (Adapted from Toyota Motors sales brochure.)

becomes too lean, the amount of unburned hydrocarbon increases again because the fuel molecules are so diluted by the air that some of them never burn.

The designer of a car has a complex set of choices to make when designing and optimizing the engine. If the goal is to reduce fuel use, then one is forced to accept high NO_x emissions. If one is designing for power and torque, then one must live with higher hydrocarbon and CO emissions. The compromise position around 14.7 percent is just that, a compromise.

Lean-burn engines as envisioned by the auto industry would increase the air/fuel ratio to as high as twenty, where the production of nitrogen oxides is minimized. Note that the resulting engine will require more gasoline because the efficiency is reduced. It will get fewer miles per gallon than a car running a richer mixture. The technology in the lean-burn engine is designed to get more of the fuel to burn using improved engine design.

Catalysts and the Catalytic Converter

The method used in the United States to remove pollutants from car exhausts is the ternary catalytic converter. The purpose of this device is to remove the NO_x, CO, and unburned hydrocarbons after they leave the engine and before they are released into the environment via the car's tailpipe. Both CO and hydrocarbons can be removed by adding more air to complete the combustion. Removing NO_x is more complicated because the reactions that eliminate NO_x are normally very slow. All the reactions required to clean the exhaust gases are too slow under normal conditions, so the device known as a catalytic converter is needed.

The study of how fast chemical reactions happen is known as chemical kinetics. The first discovery in this field was that the higher the concentration of reacting chemicals, the faster the reaction proceeds. This means that as the concentration of reacting molecules decreases, so does the rate of reaction. (This is part of the problem with the lean-burn engine.) Reactions can also be speeded up by increasing the temperature. The other way to speed up a chemical reaction is to add a substance known as a *catalyst*, which makes the reaction go faster but does not actually get used up in the reaction. Because it is not consumed, only

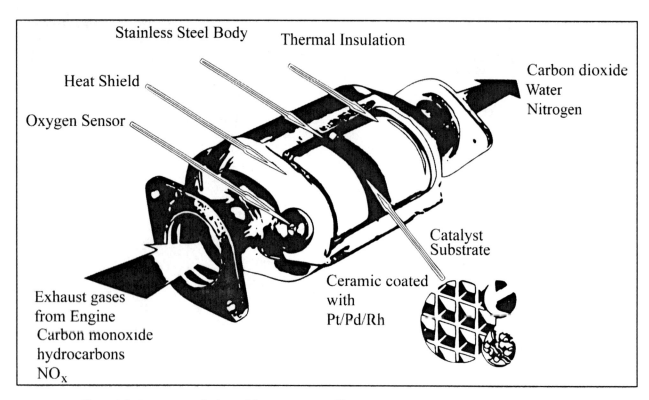

FIGURE 23 The catalytic converter. (Adapted from Master Muffler & Brake, Utah, www.mastermuffler.net/wp-content/uploads/2012/02/4551977541.jpg.)

small amounts of the catalyst are normally required.

Figure 23 shows a diagram of a catalytic converter. The manufacture of these involves using a honeycomb ceramic substrate that is porous and allows the exhaust gas to flow through readily. This is coated with a very thin layer of the actual catalytic chemical.

The catalyst is a very small amount of a mixture of platinum, palladium, and rhodium. These metals are rare and expensive, but the amount used is small. The costs of the metal to make a catalytic converter was between $100 and $200 in 1987.

The catalytic converter in Figure 23 has two separate zones in which different reactions occur.[2] The exhaust gas enters the first chamber where NO_x

is decomposed to N_2 and O_2. The oxygen recovered from the NO_x is then used to combine with the CO and hydrocarbons in the second chamber to make water and CO_2. The oxygen sensor is used to ensure that there is enough oxygen in the exhaust to complete the reaction. One of the complications of using catalytic converters is the need to adjust the engine so that the correct mixture of gases reaches the catalytic converter. Thus, the control systems are a critical component of making the overall engine system work properly to achieve maximal reduction of all three pollutants.

Units of Concentration—ppm, ppb, µg/dL, and µg/m³

For the discussion of an air pollutants like NO_x, the most commonly used term for concentration is parts per million (ppm) or parts per billion (ppb). You are certainly familiar with the use of percent

---

2. This is called a *ternary catalytic converter* because it removes three kinds of pollutants.

(%). A percentage can be calculated from the ratio of any two quantities having the same units. The resulting ratio is a dimensionless[3] ratio, which is converted to a percentage by multiplying the ratio by 100. One can also think of percent as part per hundred or pph. Air contains about 20 percent oxygen. This means that if you have 100 molecules of gas in the air, 20 will be oxygen and the rest will be other gases, mostly nitrogen with a little bit of argon, CO_2, and water vapor. Here, 20 percent means 20 parts oxygen per 100 parts air. We can think of ppm as one part in a million parts. So an ozone concentration of 100 ppm would be interpreted as 100 molecules of ozone in 1 million molecules of air.

Parts per billion (ppb) is the same type of ratio but is used for pollutants at even lower concentration. So a pollutant present at 50 ppb would have 50 molecules of pollutant for every billion molecules of air. This may not seem like a lot, but this is the range in which ozone normally occurs.

The other unit for reporting air pollution is micrograms per cubic meter (µg/m³). Converting this to ppm or ppb is a bit complicated and requires that you know the molecular weight of the pollutant and the temperature of the air. Calculators to make this conversion are available on the Internet. For most purposes, the equations here will give a reasonable approximation. The molecular weight of ozone is 48.0 g/mol. The number 24.45 is a conversion factor.

> ppm concentration = mg/m³ × 24.45 × molecular weight of pollutant
>
> ppb concentration = µg/m³ × 24.45 × molecular weight of pollutant

Thus, 50 µg/m³ of ozone corresponds to about 25 ppb ozone.

---

3. Dimensionless numbers result from the ratio of two values with the same units.

Concentrations in solution are also reported in ppm and ppb. For these cases, the ratio in weight of the pollutant is divided by the weight of the solution. Because 1 milliliter (mL) of water weighs 1 gram, it is common to use weight and volume interchangeably, even though this is not always strictly true.

> 1 ppm = 1 g pollutant/1 million g solution = 1 g pollutant/1,000 L = 1 mg/L = 1 µg/mL
>
> 1 ppb is 1,000 times less, or 1 µg/L, or 1 ng/mL

The clinical analysis of lead in blood by clinical tradition uses a different unit: micrograms/deciliter (µg/dL). One dL is 100 mL. The common action level for lead in blood is 10 µg/dL, which corresponds to 100 ppb.

Lead Pollution and NO_x

The relation of lead to NO_x lies in the fact that the catalytic converters used to remove NO_x are quickly destroyed by the presence of lead in gasoline. As little as one tankful of leaded gas can destroy a catalytic converter that costs $400 to $800 to replace. Therefore, a policy to require catalytic converters must be accompanied by a policy to provide unleaded gas for the people who buy the cars. Because Europeans travel widely on holiday, there would be serious consequences for tourism if any country failed to provide unleaded gasoline. And any country that requires catalytic converters on their cars must ensure a supply of unleaded gas or the catalytic converters will be destroyed every time people travel.

It is important to note that no one should propose adding catalytic converters to existing cars as part of a treaty. The debate is only about adding them to new cars. Cars are used for an average of about 7 years in western Europe, so this would lead to a transition time of about a decade, after which the vast majority of all cars would have catalytic converters. This transition could be accelerated by offering incentives to replace older cars with new ones.

If catalytic converters are required, a side benefit of the use of unleaded gasoline will be a dramatic reduction in lead pollution in the environment. Our discussion of environmental lead is drawn from several publications that are referenced in the sources for the game. One publication is a report from *Science* that details the extent of lead pollution and the history of the use of this toxic metal.[4] This article makes it clear that leaded gasoline is by far the largest source of environmental exposure to lead. The others from the *Atlantic Monthly* and the *Nation* detail how lead came to be added to gasoline.

Leaded Gasoline

From around 1912 until 1922, Charles Kettering and his engineers at Dayton Engineering Laboratories Company (DELCO) in Ohio worked on the problem of how to prevent preignition in engines. Kettering sold his business to General Motors in 1919 after he invented the electric starter for cars, and he continued to work for General Motors in his Dayton laboratories. Kettering was concerned that there was not enough oil to support the growing auto industry. Also, a number of substances were mixed with gasoline in the search for a way to stop knocking. A successful additive would allow more high-quality gasoline to be produced from a barrel of oil. It would also allow engines to be more efficient and to operate at higher compression ratios, giving more power as well as more efficiency. Possibly the best answer, and the one supported by the likes of Thomas Edison and Henry Ford, was to use ethyl alcohol (ethanol) in amounts from 10 to 30 percent as a fuel additive.

A side benefit of ethanol was that it reduced the smelly emissions from cars, but it had several practical and economic problems. Ethanol had to be added in large amounts, meaning that a large supply would be required. It was also something that anyone could brew up in their backyard, which the oil industry hated. On the other hand, the U.S. Department of Agriculture was in favor of ethanol as a way to support farmers who would be growing the biofuel. The oil companies had the technology to thermally crack petroleum to get more high-quality fuel, but it required major capital investment and would actually result in consumers needing less gas because it was more efficient.

In 1921, Thomas Midgley Jr., an engineer in Kettering's laboratory, discovered that tetraethyl lead (TEL), a chemical first made in 1854, was a powerful antiknock agent. Unfortunately, TEL was a highly poisonous chemical—almost immediately, people began to question the wisdom of using TEL. It was pointed out that for every gallon of gasoline a car burned, it would emit 4 grams of lead out the tailpipe. There were also concerns over whether TEL could be safely manufactured.

The impetus to use TEL was purely financial on the part of General Motors. Pierre DuPont, head of the wealthy family who had purchased a majority stake in General Motors, selected Alfred Sloan as the head of General Motors. The DuPont family also owned major stakes in Standard Oil. TEL was a product that could be patented, which meant it could become a profit center for both DuPont Chemical and Standard Oil. Kettering built the first production plant for TEL in Dayton, Ohio, and began supplying it to Standard Oil. DuPont built a second plant in New Jersey. General Motors and Standard Oil then built a third plant in New Jersey. All three plants quickly had workers die from lead poisoning. The worst of these was in the Standard Oil plant in Bayway, New Jersey: five workers died very soon after the plant began operation, and thirty-five others suffered severe lead poisoning. Ultimately, 80 percent of the workers at Bayway died or were seriously poisoned.

The risks of TEL's manufacture and the emissions of lead from car exhausts prompted a few scientists and the U.S. Surgeon General to call for research into the risks of using TEL, but their

---

4. Settle and Patterson, "Lead in Albacore," 1167–76.

efforts were blocked by government officials with connections to industry. The government decided to accept the word of the industry, and no real studies were conducted on the safety of leaded gas for decades. TEL was branded as "Ethyl" to obscure from the general public the fact that it contained lead. An interesting anecdote from this period was that Thomas Midgley, who had discovered the effect of TEL on knocking, traveled the country touting the safety of TEL by pouring the clear liquid on his hands. Periodically, he would need to retire from the tour to recover from lead poisoning.

In addition to TEL providing large profits for General Motors and DuPont, the availability of high-octane gasoline allowed General Motors to build more powerful and efficient cars and to overtake Ford as the largest auto company. Needless to say, Kettering and Sloan became wealthy, and the Memorial Sloan-Kettering Cancer Center owes its name and its existence to them.

There are many ways to manipulate the octane rating of gasoline beside adding ethanol or TEL. Gasoline also contains benzene and other aromatic hydrocarbons that change its octane rating. Refineries can use thermal cracking to introduce more branched-chain, higher-octane rating components. Other additives such as methanol and methyl tert-butyl ether also can be added to adjust the octane number.

Directly relevant to the situation in Europe in 1987 is the fact that many refineries, especially those in Eastern Europe, use essentially the same simple distillation process used in 1910 to get gasoline from crude oil. Throughout Europe in the 1980s, TEL was widely used. Even though some more technologically advanced countries had already invested in thermal crackers in refineries to get more high-quality gasoline, they would add TEL to get the maximum amount of gasoline per barrel of oil.

In 1970, the leaded gasoline in common use ranged from 90- to 100-octane. Modern unleaded fuels in the United States are typically 86- to 91-octane. Cars that used carburetors in the 1980s, when this game takes place, would need to be adjusted by a mechanic to use fuel of a different octane rating. Running a car tuned for 93-octane on 95-octane would mean that the driver was wasting some of the advantages of the higher octane fuel. Alternatively, if a driver used fuel with an octane rating of 91-octane in a car tuned for 95-octane, there would be frequent episodes of preignition under heavy driving loads. As a general rule, higher octane gasoline allows higher performance and more efficiency (more miles per gallon) for a car.

Lead is added to gasoline at levels of 0.15–0.8 g/liter. So burning a 10-gallon tank of leaded gasoline in a car can lead to emission of 5.5 to 30.0 grams of lead into the environment. The emission is primarily in the form of small particles of lead bromide. These particles can be inhaled from dust, and they are commonly picked up on the hands of children who are crawling and playing and then are transferred to their mouth when they eat. The lead also is absorbed by plants and becomes part of the food chain. Lead compounds from exhaust gases also dissolve in water and may be ingested wherever surface water is being used as the drinking water supply.

Lead in gasoline also causes a number of problems for car parts. Shortly after TEL began to be used, it was observed that spark plugs tended to short out or foul with metallic lead. When this happened, the spark plug would no longer work. A second additive was found to prevent fouling, ethylene dibromide (EDB). EDB works by converting lead to lead bromide, which comes out in the exhaust instead of building up in the engine. The bromide from excess EDB is converted in the exhaust system into hydrobromic acid, a strong acid that rapidly corrodes the muffler. Thus, large repair costs are associated with the use of leaded gasoline—the initial cost of the car and of the gasoline is less with leaded gasoline, but the maintenance costs of the car are greater.

The lead from TEL also reduces the heat in the engine and protects the valves from corrosion by the hot exhaust gases going out, which allowed less expensive valve materials to be used for the valves. Because most cars on the road in 1986 were built with valves designed for leaded gasoline, using unleaded gasoline in them could cause valve damage. However, the amount of lead required to protect valves is typically less than the amount found in most leaded gasoline. Also, leaded gasoline has a higher octane rating than unleaded gasoline, and cars designed with high-compression engines need these high octane ratings to work properly. (The author's 1970 MGB required 95-octane gasoline to operate properly, a level no longer available in the United States.)

Lead in the Environment

The history of the use of lead and its prevalence in the environment has been studied in detail by Settle and Patterson in 1980.[5] Their study used various methods, including analysis of glacial ice deposited thousands of years ago, to determine the level of environmental lead before the beginnings of human industry. The total use of lead has increased dramatically over time as shown in the Figure 24. Note that the vertical scale on the figure is in powers of 10. Total emissions have increased by more than a factor of 1 million (10^6) over the 5,000 years of human history.

Previous studies of lead in modern and prehistoric Americans have been erroneous due to errors caused by the high background levels of lead in the modern environment. Their study concluded that a 10,000-fold increase in environmental background lead has occurred since prehistoric times, which makes accurate analysis difficult for all but the most careful laboratories. Settle and Patterson showed that most results by government laboratories are too high by up to a factor of 1,000, leading to erroneous conclusions about the natural background amounts of lead in the environment and food. They were able to demonstrate that the vast majority of lead in the contemporary environ-

5. Ibid.

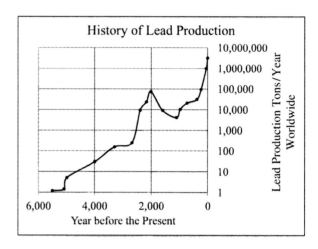

FIGURE 24 World lead production during the past 5,500 years. (Data from Settle and Patterson, 1980.)

ment is a result of human influence. We are exposed to lead in the air we breathe, the water we drink, and the food we eat at levels known to produce negative health consequences.

Human pollution of the environment with lead began over 5,000 years ago when people first learned to take sulfide ores, roast them, and obtain lead/silver alloys. The process of *cupellation* is used to separate the valuable silver from the lead; because up to 400 times more lead than silver occurs naturally in ores, the process produces a large amount of lead. Because lead has a low melting point and is very malleable, it found a number of uses. Settle and Patterson estimated that from 5,000 years ago up to the introduction of coinage around 2,700 years ago, human production of lead averaged about 160 tons per year. This amount rose almost 100-fold with the introduction of coinage, to possibly 10,000 tons per year. They estimated that lead production in the Roman Empire rose to 80,000 tons per year. Production rose again with the Industrial Revolution, from around 100,000 tons in 1700 to 1 million tons in 1920 and to 3 million tons in 1980.

This dramatic increase in lead production has resulted in human exposure to lead in both air and water, which in turn has led to poor analyses and faulty assumptions about the natural background level. Settle and Patterson estimated the natural background level to be only 0.04 ng/m^3. By contrast, the lead concentration in the ambient air in North America ranges from 500 to 10,000 ng/m^3. Thus, human exposure to lead is 10,000 to 200,000 times higher than prehistoric exposure.

Similar results are found for pollution of water. The U.S. Environmental Protection Agency's limit for lead in drinking water is 50 parts per billion (ppb). Settle and Patterson estimated the natural background value to be <0.02 ppb or 20 parts per trillion: 2,500 times *less* than what is now allowed in drinking water. The erroneous high values obtained by most laboratories have led people to believe that 50 ppb was close to a natural level when it is clearly not, based on Settle and Patterson's findings. Analytical methods to conduct analysis at the 0.02 ppb level do not exist, making it difficult to enforce a level in water that would be close to the natural, preindustrial level.

The conclusions by Settle and Patterson about the sources of lead exposure for prehistoric and modern people are shown in Table 16. The sources of lead in the modern environment and the extent to which they influence people are detailed in Table 17.

The total quantity of lead produced by a particular source is only one factor. If that were the only consideration, then coal burning and sea spray would be the two primary sources of lead. But these total values must be adjusted for the amount of total lead that is actually emitted into the environment. The emission factor is the fraction of the total lead that enters the environment in a way that humans can be exposed to it. The values in the table are in grams per kilogram, so a source that was 100 percent emitted would have an emission factor of 1,000 g/kg. You can see that 70 percent of the lead from leaded gasoline is emitted into the environment. Even though the total production of lead alkyls (tetraethyl lead and similar compounds) is a tiny fraction of total lead production, it represents 70 percent of the total lead emitted into the environment every year.

TABLE 16 Comparison of prehistoric and modern lead exposure per person

Source	Prehistoric Natural (nanograms)	1980 Urban American (nanograms)
Air	0.3	6,400
Water	<2	1,500
Food	<210	21,000
Total	<210	29,000

Source: Settle and Patterson, 1980, p. 1173.

Lead and Health

Settle and Patterson concluded that modern exposure to lead is 100 to 1,000 times greater than in prehistoric times. This results in high levels of lead in the blood and the conclusion that the average American is suffering from subacute lead toxicity. An increase of only five times the current level would produce acute lead toxicity.

Lead in the body travels with calcium and barium in biochemical processes. The effects of lead have been studied, but there is no way to study normal biochemical systems where the lead levels are as low as prehistoric natural levels. That is, lead in the environment and in the laboratory is so high and widespread that it is currently impossible to know how a normal biochemical system acts in the absence of lead. As a result, the true effects of lead are underestimated. We do know that lead interferes with oxygen transport by heme molecules in the blood, leading to anemia. It also interferes with neurological processes, leading to mental impairment. It is safe to conclude that all modern recommendations of a safe level of lead are dramatically higher than they should be.

By the 1970s, data showed that the average U.S. citizen was ingesting about 30 μg/day of lead. The

TABLE 17 Production and emission of lead

Source	Production (billion kg/year)	Emission Factor (g/kg)	Lead Emissions (thousand kg/year)
Natural			
Windblown and volcanic dust	200	1×10^{-2}	2,000
Salt spray	1,000	$<1 \times 10^{-7}$	<1,000
Forest foliage	100	$<1 \times 10^{-5}$	<100
Volcanic sulfur	6	2×10^{-4}	1
Total			2,000
Industrial			
Leaded gasoline	0.4	700	280,000
Iron smelting	780	0.06	47,000
Lead smelting	4	6	24,000
Zinc and copper smelting	15	2.8	42,000
Coal burning	3,300	4.5×10^{-3}	15,000
Total			400,000

Source: Settle and Patterson, 1980, p. 1171.

quantity required for acute toxicity is 150 µg/day. This means that the average American was within a factor of 5 of acute lead poisoning. As the data in Table 18 show, in the 1970s the vast majority of young children in the United States had blood levels above 10 µg/dL, with a mean value of 15 µg/dL. The same Centers for Disease Control and Prevention report suggested that a blood lead level of 10 µg/dL was associated with a 7 point decline in IQ and well as other physical, behavioral, and developmental problems.

The value of 10 µg/dL was chosen as an action level in the United States, and anyone with a blood concentration above this level should be treated to reduce lead exposure. However, no safe level of lead in the blood has been established, and any lead is considered a bad thing. As noted by Settle and Patterson, the preindustrial levels of blood lead were far less than the current values. Levels above 20 µg/dL typically require action to reduce exposure, and levels above 45 µg/dL are typically

TABLE 18 Lead ingestion in the United States in 1970s

Year	Percentage with BLL>10 (µg/dL)	Geometric Mean BLL (µg/dL)
1976–80	88.2	15.0

Source: Centers for Disease Control and Prevention (CDC) Advisory Committee on Childhood Lead Poisoning Prevention (2007), table 1.
Note: BLL is blood lead level.

treated medically. Above about 70 μg/dL an individual will show signs of acute lead poisoning and require immediate treatment.

Large numbers of children, especially in inner city areas with heavy traffic, suffer from chronic and even acute lead poisoning. The symptoms of this are anemia, learning disorders, and behavioral problems. Children with high lead levels have lower IQ, drop out of school more often, and are more likely to become criminals. Thus, lead pollution places a heavy cost on society, including the costs of incarcerating criminals and the lost opportunities in a population with decreased intelligence potentially contributing less to society.

Blood lead levels in Europe have also been studied. The World Health Organization (WHO) found that in the mid-1980s the worst contamination was found in Eastern Europe, where the mean blood levels were as high as 50 μg/dL, found in one city in Ukraine. The blood levels in Poland and Hungary were lower, averaging around 5 μg/dL in the cities studied. Values in Germany averaged 7.4 μg/dL, and France averaged 11 μg/dL. One study in this time frame indicated that children in Sweden averaged 5.3 μg/dL.[6] Data from Great Britain suggested an average value around 6 μg/dL.[7]

The data from European studies has shown generally lower lead levels in western Europe than in the United States. The greater use of passenger cars in the United States can be assumed to be the cause of this difference. Lead in most countries is less than the 10 μg/dL action level, but it is widely understood that there is no safe level of lead in the blood.

Readings on Lead

Additional detail about environmental lead can be found in the following publications.

---

6. Rudnai, "Levels of Lead in Children's Blood."
7. O'Donohoe, et al., "Blood Lead in U.K. Children," 219–23.

Centers for Disease Control and Prevention (CDC) Advisory Committee on Childhood Lead Poisoning Prevention, "Interpreting and Managing Blood Lead Levels <10 μg/dL in Children and Reducing Childhood Exposures to Lead," *MMWR. Recommendations and Reports: Morbidity and Mortality Weekly Report* 56, no. RR-8 (2007): 1–16; errata in *MMWR. Morbidity and Mortality Weekly Report* 56, no. 47 (2007): 1241.

K. L. Kitman, "The Secret History of Lead," *Nation*, March 20, 2003, www.thenation.com/article/secret-history-lead.

M. Lovei, *Phasing Out Lead from Gasoline: Worldwide Experience and Policy Implications*, World Bank Technical Paper No. 397, Pollution Management Series. Washington, D.C.: World Bank, 1998, http://siteresources.worldbank.org/INTURBANTRANSPORT/Resources/b09phasing.pdf.

H. L. Needleman, "The Removal of Lead from Gasoline: Historical and Personal Reflections," *Environmental Research Section A* 84 (2000): 20–35.

J. O'Donohoe, S. Chalkley, J. Richmond, and D. Barltrop, "Blood Lead in U.K. Children—Time for a Lower Action Level?," *Clinical Science* (London) 95 (1998): 219–23.

D. Owen, "Octane and Knock," *Atlantic* 259 (1987): 53–61.

P. Rudnai, "Levels of Lead in Children's Blood," Fact Sheet 4.5, RPG$_Chem_Ex1 (World Health Organization/European Environment and Health Information System, December 2009), www.euro.who.int/__data/assets/pdf_file/0003/97050/4.5.-Levels-of-lead-in-childrens-blood-EDITING_layouted.pdf.

D. M. Settle and C. C. Patterson, "Lead in Albacore: A Guide to Lead Pollution in Americans," *Science* 207, no. 4436 (1980): 1167–76.

OPTIONAL TEXTS ON ECOLOGY

J. Lovelock, *Gaia: A New Look at Life on Earth*, rev. ed. (New York: Oxford University Press, 2000; first published in 1979).

A. Leopold, *A Sand County Almanac, and Sketches Here and There* (Oxford: Oxford University Press, 1949).

B. Devall and G. Sessions, *Deep Ecology: Living as If Nature Mattered* (Layton, Utah: Gibbs M. Smith, 1985); see especially chapters 2-4.

WEBSITES WITH ENVIRONMENTAL DATA

The Norwegian Meteorological Institute's site contains reports beginning in 1981. This is a particularly good resource for the Scandinavian faction, but can be used by everyone.

www.emep.int/mscw/mscw_publications.html#1985

A number of websites provide pollution and economic data. They may be slightly useful, but remember that once the game begins, the decisions made in the class will determine the course of history. Therefore, you cannot use any post-1989 economic or pollution data that you find in databases. The Convention on Long-Range Transboundary Air Pollution website's Emissions Database lets you to enter any pollutant and year and see a Google Map or Google Earth view with an overlay of various pollution parameters. (Note that you must install Google Earth to use this feature.)

www.ceip.at/webdab-emission-database/

Data from after the year of the game should not be used in your arguments. Here are more specific instructions:

- Select WebDab on the left side, and choose how you want the data displayed. The Google Maps option is for visualizing the data.
- Click on the Search for Gridded Emissions icon.
- Select the country or area and year, and click on Search.
- Select the pollutant you wish (such as SO_2), then click on Search.
- Select the sources on the next screen, and click on Search to download a Google Earth file.
- Open the file, and Google Earth will display the data.

The Reacting Consortium Library's Acid Rain Game Simulator lets you to prepare graphs showing a number of variables relevant to your arguments. These graphs can display a number of variables at the same time and also show the potential benefits and costs of pollution reductions from 0-50 percent.

http://reactingconsortiumlibrary.org/node/91

The purpose of this simulator is to allow you to think about how different countries compare in size, wealth, pollution levels, and energy use. You should try to make at least one graph that supports the arguments of your country or faction to use during the game.

Figure 25 is a simplified an example of what this tool can display. The horizontal axis shows national gross domestic product, and the vertical axis shows the total population. Thus, the large, wealthy countries are in the top right and the poor, small countries in the lower left. The size of the circles shows the SO_2 emissions per capita for the countries. The circles are shaded to show the total SO_2 emissions. So the United Kingdom and West Germany are at the top right, and Ireland is at the lower left. The circle for the United Kingdom is smaller than West Germany's, but the total emissions are larger for the United Kingdom so the circle is darker. And Sweden is small but relatively rich. This graph might be used by Ireland or Sweden to argue that they are not causing the problem and do not need to do anything. It also highlights that Czechoslovakia has a small, poor population but is a very large polluter.

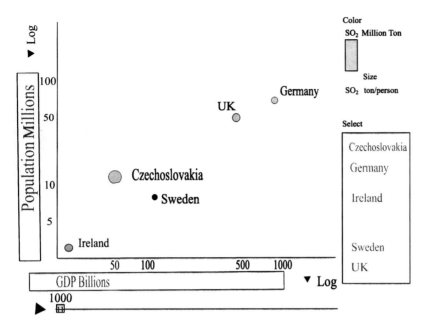

FIGURE 25 Example of game simulator tool.

There is a bar across the bottom with a slider, here shown as a grey box on the line to the right of the large black triangle. Dragging the bar across the bottom of the graph on the website shows how things change when one goes from taking no action to reducing pollution by 50 percent. Each axis can be changed to display either a linear scale or a logarithmic scale. Clicking on the small triangle to the right of the Log indicator brings up a box to make the change. The boxes containing the words "BTU" and "Population" also open to allow selection of any of the categories of data to display on the axis. The "Color" and "Size" boxes open to let you select a category of data for the color and size. In the example, total SO_2 was selected for color and SO_2 per person for size. Finally, the box with the names of countries allows you to select which countries are labeled. All of the countries will show up on the graph but only the selected ones are labeled. This allows you to highlight the countries you are interested in comparing. Several countries are omitted in the box for clarity in this figure. Experiment with the calculator and try to find a visual display that makes the point your country wants to make.

Bibliography

Abrahamsen, G., A. O. Stuanes, and B. Tveite. "Effects of Long Range Transported Air Pollutants in Scandinavia." *Water Quality Bulletin* 8 (1983): 89-95, 109.

Allison, L. *Ecology and Utility: The Philosophical Dilemmas of Planetary Management*. Rutherford, N.J.: Fairleigh Dickinson University Press, 1991.

Amann, M., D. Derwent, B. Forsberg, O. Hänninen, F. Hurley, M. Krzyzanowski, F. de Leeuw, et al. *Health Risks of Ozone from Long-Range Transboundary Air Pollution*. Copenhagen: World Health Organization Regional Office for Europe, 2008.

Andonoiva, L. B. *Transnational Politics of the Environment*. Boston: MIT Press, 2004.

Aristotle. "Politics," in *Aristotle in 23 Volumes*. Vol. 21. Translated by H. Rackham. Cambridge, Mass.: Harvard University Press, 1944.

Argyrou, V. *The Logic of Environmentalism: Anthropology, Ecology and Postcoloniality*. New York: Berghahn Books, 2005.

Barrett, M., and R. Protheroe. *Sulphur Emission from Large Point Sources in Europe*. 2nd ed. Göteborg: Swedish NGO Secretariat on Acid Rain, 1994. www.airclim.org/sites/default/files/documents/APC3.pdf.

Beyers, R. J. "The Metabolism of Twelve Aquatic Laboratory Micro Ecosystems." *Ecological Monographs* 33 (1963): 281-305.

———. "The Microcosm Approach to Ecosystem Biology." *American Biology Teacher* 26 (1964): 491-98.

Boehmer-Christiansen, S., and J. Skea. *Acid Politics: Environmental and Energy Policies in Britain and Germany*. New ed. London: Belhaven Press, 1993.

Boyle, P. F., J. W. Ross, J. C. Synnott, and C. L. James. "A Simple Method to Measure the pH Accurately in Rain Samples." In *Impact of Acid Rain and Deposition on Aquatic Biological Systems, ASTM STP 928*, edited by B. G. Isom, S. D. Dennis, and J. M. Bates, 98-106. Philadelphia: American Society for Testing and Materials, 1986.

Brower, J. E., and J. H. Zar. *Field Laboratory Methods for General Biology*. Dubuque, Iowa: William C. Brown, 1977.

Carson, R. *Silent Spring*. New York: Houghton Mifflin, 1962.

Cooke, G. D. "The Pattern of Autotrophic Succession in Laboratory Microcosms." *BioScience* 17 (1967): 717-21.

Costanza, R., R. d'Arge, R. de Groot, S. Farber, M. Grasso, B. Hannon, K. Limburg, et al. "The Value of the World's Ecosystem Services and Natural Capital." *Nature* 387 (1997): 253-60.

Devall, B., and G. Sessions. *Deep Ecology: Living as If Nature Mattered*. Layton, Utah: Gibbs M. Smith, 1985.

European Environment Agency. *European Union Emission Inventory Report 1990-2012 under the UNECE Convention on Long-range Transboundary Air Pollution (LRTAP)*. EEA Technical Report. No. 12/2014. Luxembourg: Publications Office of the European Union, 2014. www.eea.europa.eu/publications/lrtap-2014.

Feister, U., U. Pedersen, E. Schulz, and S. Hechler. *Report No. 2 Ozone Measurements January 1986-December 1989*. EMEP/CC Report 8/90 (September 1990).

Ferens, M. C., and R. J. Beyers. "Studies of a Simple Laboratory Micro Ecosystem: Effects of Stress." *Ecology* 53 (1972): 709-13.

Fisher, B. E. A. "Deposition of Sulphur and the Acidity of Precipitation over Ireland." *Atmospheric Environment* 16, no. 11 (1982): 2725-34.

Glacken, C. J. *Traces on a Rhodian Shore: Nature and Culture in Western Thought from Ancient Times to the End of the Eighteenth Century.* Berkeley: University of California Press, 1967.

Gorden, R. W., R. J. Beyers, E. P. Odum, and E. G. Egon. "Studies of a Simple Laboratory Micro Ecosystem: Bacterial Activities in a Heterotrophic Succession." *Ecology* 50 (1986): 86–100.

Gryparis, A., B. Forsberg, K. Katsouyanni, A. Analitis, G. Touloumi, J. Schwartz, E. Samoli, et al. "Acute Effects of Ozone on Mortality from the 'Air Pollution and Health: A European Approach' Project." *American Journal of Respiratory and Critical Care Medicine* 170 (2004): 1080–87.

Hasselmann, K. "Are We Seeing Global Warming?" *Science* 276 (1997): 914–15.

Henderson, D. E. *Air Pollution and Risk Analysis.* NLA Monograph Series No. 13. Stony Brook: State University of New York, 1990.

Herendeen, R. A. *Ecological Numeracy: Quantitative Analysis of Environmental Issues.* New York: John Wiley and Sons, 1998.

Iversen, T., J. Saltbones, H. Sandnes, A. Eliassen, and O. Hov. *Airborne Transboundary Transport of Sulphur and Nitrogen over Europe-Model Descriptions and Calculations.* EMEP MSC-W Report (August 1989).

Jacks, G., G. Knutsson, L. Maxe, and A. Fylkner. "Effect of Acid Rain on Soil and Groundwater in Sweden." *Ecological Studies: Analysis and Synthesis* 47 (1984): 94–114.

Kerr, R. A. "Greenhouse Forecasting Still Cloudy." *Science* 276 (1997): 1040–42.

Leopold, A. *A Sand County Almanac, and Sketches Here and There.* Oxford: Oxford University Press, 1989. First published in 1949.

Likens, G. E. "An Experimental Approach to the Study of Ecosystems." *Journal of Ecology* 73 (1985): 381–96.

Likens, G. E., C. T. Driscoll, and D. C. Buso. "Long-Term Effects of Acid Rain: Response and Recovery of a Forest Ecosystem." *Science* 272 (1996): 244–45.

Lovelock, J. *Gaia: A New Look at Life on Earth*, rev. ed. New York: Oxford University Press, 2000. First published in 1979.

Marsh, G. P. *Man and Nature; or, Physical Geography as Modified by Human Action.* New York: Charles Scribner, 1864.

Maugh, T. H., II. "Acid Rain's Effects on People Assessed." *Science* 226 (1984): 1408–10.

Meszaros, E. "Les pluies acides en Hongrie" [Acid Rain in Hungary]. *Pollution Atmospherique* 110 (1986): 112–15.

Naess, A. *Life's Philosophy: Reason and Feeling in a Deeper World.* Athens: University of Georgia Press, 2002.

Odum, H. T., and C. M. Hoskin. "Metabolism of a Laboratory Stream Microcosm." *Publications of the Institute of Marine Science* [University of Texas] 4 (1957): 115–33.

Office of Air Quality Planning and Standards. *National Air Pollutant Emission Trends, 1900–1995.* Report No. EPA-454/R-96-007. Research Triangle Park, N.C.: U.S. Environmental Protection Agency, 1996. Available at https://web.archive.org/web/19970801103927/http://www.epa.gov:80/oar/emtrnd/report96.htm.

Office of Technology Assessment. *Acid Rain and Transported Air Pollutants: Implications for Public Policy.* Report OTA-O-204. Washington, D.C.: U.S. Government Printing Office, 1984. Available at www.princeton.edu/~ota/disk3/1984/8401_n.html.

Organisation for Economic Co-operation and Development. *The Costs and Benefits of Sulfur Oxide Control: A Methodological Study.* Paris: OECD, 1981.

———. *The OECD Programme on Long Range Transport on Air Pollutants: Measurements and Findings.* Paris: Organisation for Economic Co-operation and Development, 1977.

———. *The OECD Programme on Long Range Transport of Air Pollutants: Summary Report.*

Paris: Organisation for Economic Co-operation and Development, 1977.

O'Sullivan, D. A. "European Concern about Acid Rain Is Growing." *Chemical and Engineering News* 63 (1985): 12–19.

Paces, T. "Sources of Acidification in Central Europe Estimated from Elemental Budgets in Small Basins." *Nature* 315 (1985): 31–36.

Pantani F., E. Barbolani, S. Del Panta, and F. Bussotti F. "Rilavemento di piogge acide in comprensori della Toscano." *Maggio- Giugno* 35 (1984): 135–41.

Park, Chris C. *Acid Rain: Rhetoric and Reality*. London: Methuen, 1987.

Perrone, P. A., and Gant, J. R. "Indirect Photometric Detection in Anion Chromatography." *Research and Development* 26 (1984): 96–100.

Plato. "Timeus." In *Plato in Twelve Volumes*. Vol. 9. Translated by W. R. M. Lamb. Cambridge, Mass.: Harvard University Press, 1925.

Roberts, T. M., N. M. Darrall, and P. Lane. "Effects of Gaseous Pollutants on Agriculture and Forestry in the UK." *Advances in Applied Biology* 9 (1983): 1–142.

Roush, W. "Putting a Price Tag on Nature's Bounty." *Science* 276 (1997): 1029.

Ryle, M. "Economics of Alternative Energy Sources." *Nature* 267 (1977): 111–17.

Schwartz, S. E. "Acid Deposition: Unraveling Regional Phenomena." *Science* 243 (1989): 753–63.

Settle, D. M., and C. C. Patterson. "Lead in Albacore: Guide to Lead Pollution in Americans," *Science* 207, no. 4436 (1980): 1167–76.

Smidt, Stefan. *Investigations about the Occurrence of Acid Precipitation in Austria*. Mitteilungen der Forstlichen Bundesversuchsanstalt Wien 150. Vienna: Österreichischer Agrarverlag, 1983.

Toennies, G., and B. Bakay. "Photonephelometric Microdetermination of Sulfate and Organic Sulfur." *Analytical Chemistry* 25, no. 1 (1953): 160–5.

Ulrich, B., R. Mayer, and P. K. Khanna. "Chemical Changes Due to Acid Precipitation in a Loess-Derived Soil in Central Europe." *Soil Science* 130 (1980): 193–99.

Visvader, J. "Gaia and the Myths of Harmony: An Exploration of Ethical and Practical Implications." In *Scientists on Gaia*, edited by S. H. Schneider and P. J. Boston, 33–37. Cambridge, Mass.: MIT Press, 1991.

Wagner, C. K. "The Use of Aquatic Research Microecosystem in the Biology Teaching Laboratory." In *Tested Studies for Laboratory Teaching: Proceedings of the Fourth Workshop/Conference of the Association for Biology Laboratory Education (ABLE)*, edited by C. L. Harris, 59–70. Dubuque, Iowa: Kendall/Hunt, 1984.

Wetstone, G. S., and A. Rosencranz. *Acid Rain in Europe and North America: National Responses to an International Problem*. Washington, D.C.: Environmental Law Institute, 1983.

Wiseman, P. "Employers Get Pickier." AP/AOL Finance, June 26, 2013. www.aol.com/2013/06/26/top-employers-pickier-than-ever/.

World Bank. "Databank." Accessed April 4, 2007. http://databank.worldbank.org/data/home.aspx.

Acknowledgments

This game was first conceived during the authors' sabbatical leaves while studying the effects of machine-made snow on acidification of streams in Vermont. We attended a conference on acid precipitation held at Connecticut College where Prof. Lilian Andanova presented a paper describing the efforts in Eastern Europe to control pollution, which planted the seeds for this game. The combination of politics and science was irresistible. The initial outline for the game was prepared on a train ride to Washington, D.C., in April 2005. We were attending a conference with Mark Carnes at the American Association of Colleges and of Universities at Bethesda, Maryland. The purpose of the session was to help participants try out one of the established games.

Some of the background information in this document is adapted from "The Acid Rain Project—A Discovery Experiment for General Chemistry and Biology" (David E. Henderson, Michael O'Donnell, and Rebecca Thomas).

Financial support for the authors to develop this game was provided through a grant from the Mellon Foundation to Trinity College and by a grant from the Project on Secular Traditions and the Liberal Arts, Institute for the Study of Secularism in Society and Culture, Trinity College, Hartford, Connecticut. Cara Pavlak and Caitlin Farrell worked under this grant as research assistants and helped by obtaining background information, sharing their ideas for the game, and acting as readers of the draft version. The fertile discussions with Dan Blackburn, Kent Dunlop, Sean Coco, Barry Kosmin, Andrew Walsh, and Areila Keysar during the year were also much appreciated.

The encouragement of Mark Carnes in our game developments is warmly acknowledged.

We dedicate this book in memory of Prof. Michael Petterson, who was an early user of this game and served as the development editor for this project.

Appendix 1. Introduction to Environmental Philosophy

Most people never think in terms of a personal environmental philosophy, but their actions demonstrate it every day. Whether you throw your soda can in the trash or put it in the recycle bin is reflective of your environmental philosophy. The reacting games include roles that exemplify and articulate various examples of environmental philosophy, a few of which are discussed here. Although environmental philosophy is a relatively new specialty, philosophers for centuries have developed philosophical positions that relate humanity and the environment. Formal studies of environmental philosophy can illuminate our assumptions and inform our approach to environmental problems.

The earliest articulated environmental philosophy is found in the Bible. In Genesis we are told, "And God said, Let us make man in our image, after our likeness: and let them have *dominion* over the fish of the sea, and over the fowl of the air, and over the cattle, and over all the earth, and over every creeping thing that creepeth upon the earth" (Gen. 1:26 AV). The message of this passage is clear: humans are the rulers, and everything else is under their control. Like a medieval king, humans can do with the other living things as they choose.

This biblical philosophical position has continued to pervade Western thought over the millennia. René Descartes articulated this position during the Enlightenment. He viewed nature as a possession of humanity to which no rights have been granted. In the 1980s, U.S. Secretary of the Interior James Watts articulated the same philosophy for the Reagan administration. Some aides of the Reagan administration, who believed in a radical apocalyptic Christianity, were even reported to have said that we had a Christian duty to use up the resources of the Earth before the end times arrive. They could cite the parable of the talents (Matt. 25:14–30 AV) that all the resources entrusted to humanity were to be fully exploited, not just saved for later.

An alternative view of nature is also present from early history. Greek and Roman philosophers regarded the Earth as a living and even intelligent being. Plato speaks of the "soul of the world" and felt that all things had souls. Platonic philosophy became incorporated into Christianity with a creator god who was all good and who formed his creation in his own image: "He constructed this present Universe, one single Living Creature containing within itself all living creatures both mortal and immortal. And He Himself acts as the Constructor of things divine, but the structure of the mortal things He commanded His own engendered sons to execute. And they, imitating Him, on receiving the immortal principle of soul, framed around it a mortal body, and gave it all the body to be its vehicle."[1]

This position persisted well into the Middle Ages. As the seventeenth-century mathematician and astronomer Johannes Kepler wrote, "Therefore there is in the earth not only dumb, unintelligent humidity, but also an intelligent soul which begins to dance when the aspects pipe for it."[2] The idea of the Earth as a soul that creates the natural harmony was apparent to people even in early times. This philosophical position of the Earth as a single living creature returns as a scientific theory in the

---

1. Plato, "Timaeus," in *Plato in Twelve Volumes,* vol. 9, trans. W. R. M. Lamb (Cambridge, Mass.: Harvard University Press, 1925), 69c–d. Available at www.perseus.tufts.edu/hopper/text?doc=plat.+tim.

2. Quoted by J. Visvader, "Gaia and the Myths of Harmony: An Exploration of Ethical and Practical Implications," in *Scientists on Gaia,* ed. S. H. Schneider and P. J. Boston, 33–37 (Cambridge, Mass.: MIT Press, 1991), 35.

book *Gaia: A New Look at Life on Earth* (1979) by James Lovelock. The science in his book was extremely controversial in 1979, but its philosophical position about the world had ancient roots. The subsequent research prompted by *Gaia* would confirm much of what Lovelock proposed, and it became the basis of the discipline of biogeochemistry, or earth systems science.

A second ancient position was that of Aristotle. Whereas Plato believed that we could learn everything we needed to know about nature using logic, Aristotle was a proponent of observation and experiment. Aristotle would have argued that nature did nothing without a purpose. His view is more mechanical, and he suggests that the progression of nature is the result of individual beings, either consciously or unconsciously, working toward some purpose. Aristotle made extensive observations of nature and concluded:

> So that clearly we must suppose that nature also provides for them in a similar way when grown up, and that plants exist for the sake of animals and the other animals for the good of man, the domestic species both for his service and for his food, and if not all at all events most of the wild ones for the sake of his food and of his supplies of other kinds, in order that they may furnish him both with clothing and with other appliances. If therefore nature makes nothing without purpose or in vain, it follows that nature has made all the animals for the sake of men. Hence even the art of war will by nature be in a manner an art of acquisition (for the art of hunting is a part of it) that is properly employed both against wild animals and against such of mankind as though designed by nature for subjection refuse to submit to it, inasmuch as this warfare is by nature just.[3]

Aristotle's own writings leave open the question of the "Final Cause" that provides purpose and direction to Nature. As noted by Clarence Glacken, when Aristotle was adopted by Christianity, it was a simple matter to suggest that the Christian God provided the purpose and design.[4]

It is ironic that while the views of Plato and Aristotle are seen as opposing each other, they both connect to the Gaia hypothesis of Lovelock. The idea of all natural processes working together toward a stable condition called *homeostasis* for the ecosystem is directly in line with the ideas of Aristotle, and the idea of the Earth as a single living creature with a soul harkens back to both Plato and Kepler.

The idea that some elements of nature have rights of their own also has a rich history. Native Americans clearly held this as true for animals. Nathaniel Ward, in his codification of the laws of the Massachusetts Bay Colony in 1641, included protection for farm animals against abuse by their owners. None of these philosophical positions deviate from utilitarianism in that they all view humans as having the right to use anything in the environment for their own needs. The most that can be said is that they understood the responsibility not to inflict unnecessary pain in the way they used the resources around them.

The Transcendentalists of the nineteenth century, most notably Henry David Thoreau, began to reconnect with the natural world around them. Thoreau's *Walden* (1854) is a notable example of this evolving change in environmental philosophy. From the standpoint of environmental management, the most significant writing during this period was *Man and Nature* (1864) by George Perkins Marsh. (Marsh's original title for the book had been *Man the Disturber of Nature's Harmonies*.) This is the writing of a practical Vermonter,

---

3. Aristotle, "Politics," in *Aristotle in 23 Volumes*, vol. 21, trans. H. Rackham (Cambridge, Mass.: Harvard University Press, 1944), 1256b. Available at www.perseus.tufts.edu/hopper/text?doc=Perseus%3atext%3a1999.01.0058.

4. C. J. Glacken, *Traces on a Rhodian Shore: Nature and Culture in Western Thought from Ancient Times to the End of the Eighteenth Century* (Berkeley: University of California Press, 1967), 49.

not an abstract philosopher, but it marks a turning point in environmental philosophy. Marsh had traveled widely in Europe and the Mediterranean, and he had seen the impact of several thousand years of "civilization" on the landscape. At the same time, his native Vermont was being clear-cut for lumber, and the hills and mountains turned into pasture for sheep to supply the ubiquitous woolen mills in every town and village.

> Nature, left undisturbed, so fashions her territory as to give it almost unchanging permanence of form, outline, and proportion, except where shattered by geological convulsions; and in these comparatively rare cases of derangement, she sets herself at once to repair the superficial damage, and to restore, as nearly as practicable, the former aspects of her dominion.[5]
>
> Man has too long forgotten that the earth was given to him for usufruct alone, not for consumption, still less for profligate waste. Nature has provided against the absolute destruction of any of her elementary matter, the raw material of her works; the thunderbolt and the tornado, the convulsive throes of even the volcano and the earthquake, being only phenomena of decomposition and recomposition. But she has left it within the power of man irreparably to derange the combination of inorganic matter and of organic life, which through the night of eons she had been proportioning and balancing, to prepare the earth for his habitation, when, in the fullness of time, his creator should call him forth to enter into its possession.[6]

We move forward a century to the 1949 essay "The Land Ethic" by Aldo Leopold. This essay is generally considered to be one of the most important writings in the philosophical literature of the modern environmental movement. Leopold, like Marsh a century before, understood the basic cooperative nature of the ecosystem, and he speaks in the language of ethics and philosophy of the fallacies of many environmental approaches. He notes that ethics rest on the interaction of the individual with the community she inhabits. Leopold's contribution was to extend the community of humanity to include "soils, waters, plants, and animals, or collectively: the land."[7]

Leopold attacks a century of conservation efforts begun by Marsh and others as too little to be effective. He argues for the fundamental rights of all parts of the biosphere. And finally he attacks the idea that economic value is a valid way to evaluate conservation or environmental impact: "a system of conservation based solely on economic self-interest is hopelessly lopsided. It tends to ignore, and thus eventually to eliminate, many elements in the land community that lack commercial value, but that are (as far as we know) essential to its healthy functioning. It assumes, falsely, I think, that the economic parts of the biotic clock will function without the uneconomic parts. It tends to relegate to government many functions eventually too large, too complex, or too widely dispersed to be performed by government."[8] The land ethic that Leopold advocates is one in which the fabric of society internalizes the fundamental value and rights of all living things and of their interconnections. It is a society in which actions against this all-inclusive community are penalized and actions that support it are encouraged at every level, without focusing solely on the economic issues.

The ultimate expression of this position is found in the Deep Ecology movement begun by the Norwegian philosopher Arne Næss in 1972. His

---

5. G. P. Marsh, *Man and Nature; or, Physical Geography as Modified by Human Action* (New York: Charles Scribner, 1864), 29.

6. Ibid., 36.

7. A. Leopold, "The Land Ethic," in *A Sand County Almanac, and Sketches Here and There* (Oxford: Oxford University Press, [1949] 1989), 204.

8. Ibid., 214.

position grows out of the writings of Rachel Carson and Aldo Leopold. From this perspective, humanity is just one component of the larger biosphere and does not have any special position, either due to intelligence, ability to modify the environment, or the idea of a "soul." Humans are just one species of animals among many, and all species have equal right to life and resources.

Students seeking more information on this should consult *Gaia* by James Lovelock. This book was controversial when it was published in 1979. It was criticized by nonscientists as too "sciencey" and by scientists for introducing myths into a scientific treatment. Lovelock also did not use the traditional language of environmental philosophy. But Lovelock has been a central figure in the development of the environmental movement in the twentieth century, and his contributions are often not fully appreciated. He is the inventor of the electron capture detector for the gas chromatograph; without this device, the analysis of trace pesticide residues would not have been possible. The data provided by Lovelock's invention was necessary for Rachel Carson to write her seminal book *Silent Spring* (1962), which is often credited with starting the popular environmental movement in the 1970s.[9] If Rachel Carson is the mother of the environmental movement, then Lovelock is its grandfather.

Virtually every environmental science curriculum includes courses in earth systems science or geophysiology. These courses are a direct outgrowth of the ideas Lovelock originated. Lovelock took an inert, inorganic world and made it come alive. He showed the interconnections in the world and the role of biology in the inorganic environment. Every discussion of global warming includes a discussion of positive and negative feedbacks or forcing mechanisms, which are a direct outgrowth of Lovelock's science. And he introduced the most radical of all environmental philosophies, that humanity and the entire biosphere are intimately linked in a complex, interdependent community. Lovelock brings us the deep scientific understanding that underlies Leopold's land ethic. Lovelock removes humanity from a special place and dispenses with special privileges, and he suggests that the entire biosphere of Earth is in fact the highest form of life. Humans are just one part of the ecology, and if we mess things up the biosphere may grind us up. This is a radical change in thinking about the role of humanity. It is still utilitarian in nature, but it suggests that the utility that is important is not just economic advantage for people but self-preservation of Gaia.

The idea that the entire biosphere is a self-regulating organism is a double-edged philosophical sword. It can be used to argue for the need to treat Gaia as an organic being and to respect and protect all her vital functions. But the self-regulating aspect can also be interpreted as self-repairing. In this view, the biosphere can take care of itself and will adjust itself and protect itself from human excesses.

During the 1970s and 1980s, other new approaches to the relationship of humanity with the world began to emerge. Developing countries saw the poverty of their people as an environmental problem, and the growing populations of many such countries placed extreme stress on their environment. In Haiti, for example, virtually all the trees were cut for firewood for cooking due to the lack of money to buy other fuels like kerosene. This led to devastating floods and other problems. Thus, poverty and the lack of access to modern energy sources has been a greater concern for developing countries than the more esoteric issues of air pollution, pesticides, or acid rain. If there are no forests to damage, then why worry about acid rain? Developed countries, meanwhile, were focusing on issues such as acid rain and depletion of the ozone layer. This clash of interests between the wealthy, developed countries and the poor, undeveloped countries presents a huge stumbling block in formulating agreements.

---

9. R. Carson, *Silent Spring* (New York: Houghton Mifflin, 1962).

In spite of decades of development of an ecological consciousness, the majority of people involved in political and industrial leadership appear to operate from a utilitarian position, which is described in the book *Deep Ecology* as follows:

1. People are fundamentally different from all other creatures on Earth, over which they have dominion (defined as domination).
2. People are masters of their own destiny; they can choose their goals and learn to do whatever is necessary to achieve them.
3. The world is vast and thus provides unlimited opportunity for humans.
4. The history of humanity is one of progress; for every problem there is a solution, and thus progress need never cease.
5. All problems are solvable.
6. All problems are solvable by people.
7. Many problems are solvable by technology.
8. Those problems that are not solvable by technology have solutions in the social world (politics, economics, etc.).
9. When the chips are down, we will apply ourselves and work together for a solution before it is too late.[10]

Many environmentalists have worked to modify this position without changing any of the underlying assumptions. They have worked to reduce pollution and set aside land for open space. They have attempted to gain public support by publicizing environmental problems. In a sense, they have focused attention on item 8 above, to gain political support for solving problems that they feel may or may not have technological solutions. The creation of national parks and preserves is a political response to the loss of natural spaces. Urbanization has concentrated human population, which frees open space but concentrates pollution.

The idea that nature possesses rights that humans must respect is still not widely accepted. Reformers have begun to argue that higher animals have rights, though less than those of humans. The environmental movement continues its efforts to add statements of these rights to laws and to specifically limit the rights of corporations to infringe on these rights.[11]

In addition to the more practical concerns of governments, Deep Ecology challenges the prevailing philosophical positions of environmentalism in more fundamental ways. Some of these ideas are quite radical in the role they assign to humans. In general, they view humans as just one part of nature; in the more radical visions, humans are not necessarily the central or most important figure in the ecology. Vassos Argyrou characterizes these radical ecological theories as follows:

> Human beings are neither stewards of the planet nor in any ontological sense different from nature.... Human beings are denizens of the planet and a part of nature, indeed, a very small part of a wider, ongoing, possibly providential, certainly grand and extraordinary process. They are part of the process by which Life unfolds, reproduces, and propagates itself—a process that has its own logic and whose ultimate meaning human beings do not comprehend and should not interfere with.... To have thought otherwise is the mark not of improvidence but of hubris, of utter lack of respect towards something immensely larger and grander than human beings.[12]

The Deep Ecological consciousness calls for changes in lifestyle and in one's approach to the use of resources. It calls for a simpler life, one less devoted

---

10. B. Devall and G. Sessions, *Deep Ecology: Living as If Nature Mattered* (Layton, Utah: Gibbs M. Smith, 1985), 43.

11. For an example, see the Community Environmental Legal Defense Fund at www.celdf.org.

12. V. Argyrou, *The Logic of Environmentalism: Anthropology, Ecology and Postcoloniality* (New York: Berghahn Books, 2005), 50–51.

to acquisition of wealth and more to living simply in the world. In Deep Ecology, statements of Utilitarian Philosophy are rejected and replaced by these:

1. The well-being and flourishing of human and nonhuman life on Earth have intrinsic, inherent value. These values are independent of the usefulness of the nonhuman world for human purposes.
2. Richness and diversity of life forms contribute to the realization of these values and are also values in themselves.
3. Humans have no right to reduce this richness and diversity except to satisfy *vital* needs.
4. The flourishing of human life and culture is compatible with a substantial decrease of the human population. The flourishing of nonhuman life requires such a decrease.
5. Present human interference with the nonhuman world is excessive, and the situation is rapidly worsening.
6. Policies must therefore be changed. These policies affect basic economic, technological, and ideological structures. The resulting state of affairs will be deeply different from the present.
7. The ideological change is mainly that of appreciating life quality (dwelling in situations of inherent value) rather than adhering to an increasingly higher standard of living. There will be a profound awareness of the difference between big and great.
8. Those who subscribe to the foregoing points have an obligation directly or indirectly to try to implement the necessary changes.[13]

13. Devall and Sessions, *Deep Ecology*, 70.

Appendix 2. Introduction to Environmental Economics

Economics, sometimes referred to as the "dismal science," attempts to quantify economic activity and predict ways to optimize it. This game will make use of economic data in the form of gross domestic product (GDP). Although GDP is the most widely used measure of economic activity, it has a number of serious problems.

The ultimate goal of environmental economics is a cost-benefit analysis for a policy or product. The costs must include not only direct costs but also costs avoided. Thus, acid rain policy must consider not only the costs of pollution abatement but also the costs avoided—for example, avoiding the need to replace buildings or other infrastructure that will not be destroyed if acid rain is prevented.

Environmental economics provides an analysis of different approaches to reaching environmental goals in order to determine which is the most cost effective. One example is the debate over how to reduce acid rain. One approach is to legislate limits on emissions for each industry or plant. An alternative is to establish a market for emissions, set a cap for the total emissions, and allow emission rights to be bought and sold as in a stock market. The United States adopted this approach to reducing sulfur pollution in the 1980s. Many European countries are more familiar with the "command and control" approach. In part this is because European governments own much of the energy production, so it is less susceptible to market forces. The question for environmental economists is to decide which of these approaches achieves the best reduction of acid at the lowest cost.

One limitation of cost-benefit analysis is that the costs of policies are often immediate and the benefits do not appear until much later. Again using acid rain as an example, it may cost billions of dollars to put scrubbers on power plants to remove sulfur emissions. The resulting reduction in acid rain may save ten times this amount in building repairs and health care costs avoided as a result, but the savings accumulate over a period of decades and may be hard to quantify. The evaluation of future costs and benefits are adjusted to reflect their present value in a process called *discounting*.

From the point of view of those involved in environmental economics, GDP is a crude and misleading instrument for studying economic health and growth. GDP is really a measure of how money flows in the economy. Some of that flow comes from using up available resources that cannot be replaced. For example, if a country produces oil, and the production is counted in the GDP, at some point the oil will be gone. So the flow of money represents the loss of a resource that cannot be replaced, in that oil is essentially like capital that can only be spent once. Thus, a higher GDP based on consuming a limited resource is not a positive item at all—it cannot be sustained when the resource is gone.

Some of the factors that increase the GDP may make the economy look better but are actually costs. The cost of medical care due to air pollution brings more money into the economy in the form of drugs, medical bills, and hospitalization, but we would all agree that more of such costs are a detriment to society.

Yet another example of the weakness of using GDP as a measure of economic health is the case where labor that was provided with no actual cost suddenly becomes costly, as when a spouse gets a job and needs a housekeeper or babysitter. Similarly, if a product becomes scarce due to depletion and becomes more expensive, this can also increase the GDP and make the economy look better.

Economists have struggled to incorporate the value of the natural processes of the environment into their calculations of cost and benefit, or profit and loss. They struggle to assess the costs of pollution of the environment. These are called *external* costs. Because these external features are not bought and sold, there is no accurate way to determine their value. This is referred to as a "market failure" in the language of the discipline. The fact that the external costs of a process or policy are not reflected in the price leads to the failure of the market to support the most cost-effective products or policies.

The fact that the biosphere provides humanity with a variety of critical services is central to the environmentalists' position. The biosphere purifies water, produces oxygen for us to breathe, removes our waste products, transports nutrients, and keeps the temperature comfortable. Traditional economic analyses has ignored these services provided by the Earth. This has led to overuse of resources. If the limited nature of these services and resources is included in the cost of the product, then as the resource becomes scarce, the price of the product will increase. Markets are a feedback system that should prevent the overutilization of a resource if it is properly priced; unfortunately, it is not always easy to determine the true cost of these external features, given our limited knowledge of the mechanisms at work in the biosphere.

Another failure of traditional economics is the way it treats raw materials. Basic resources such as energy, water, air, and raw materials are often treated as inexhaustible resources to be used at will. Environmental economics seeks to change this by including the scarcity factor in the cost of economic activity. As with the value of the services provided by the environment, the true costs are often difficult to determine. Accurate inventories of natural resources and the cost and reliability of production are hard to obtain. For instance, crude oil is a product for which a very free global market exists. The recent instability of the price of oil is an example of the degree to which markets respond to political factors as well as to the actual amount of the resource available.

Traditionally, the value of environmental resources have been determined based on their value to people who use them either directly or indirectly. It is even possible to place a value on resources not used, such as a wilderness area which has value simply because people like to know that such an area exists. The political battle to preserve the wild areas of northern Alaska is an example. But all these values are related to humans—the value of resources to other organisms is not readily considered. Ultimately, these are issues of indirect value because the resources are connected through the cybernetic web of the biosphere. Because most aspects of the biosphere are ultimately important for humanity, they should all have value. However, when the number of connections in the chain becomes too great, we can no longer see the value and therefore fail to properly include it in the analysis. It is simple to recognize that some of the systems described by Lovelock, such as the kelp that adds iodine to the land or the plankton that produces dimethyl sulfide which in turn makes clouds to irrigate the land, would be hard to assign value. The long time constants of some of these processes is also a complication. If it takes years (or centuries) to see the impact of an activity, then it will be difficult to assign it a proper value.

Various efforts have been made to develop a "green" version of the GDP to better assess the real environmental and social development of a country. Such an index would provide more efficiency in decision making and resource use and would indicate when societies are moving backward rather than forward. When these are incorporated directly into a measure like the GDP, the result is often referred to as a "Green GDP." Examples include the Index of Social Health, the Human Development Index, and the Index of Sustainable Economic Welfare. Most of these measures indicate that the course of Western economies is at best flat, and that the appearance of growth is due to the previously discussed distortions of the GDP and the loss of fundamental stocks of resources. Critics of these results point out that the decisions of how to assign dollar value to many of these features are inherently biased.

A more detailed introduction to environmental economics can be found at the sites ELAW: Environmental Law Alliance Worldwide (www.elaw.org) and Ecosystem Valuation (www.ecosystemvaluation.org).

Appendix 3. Using Numbers to Make Arguments

It should be obvious that if one is going to do cost-benefit analysis it will be necessary to do some arithmetic to make an argument. Several spreadsheets are provided for the course, which contain pollution and economic data. These can be helpful as a starting point for various kinds of analysis, but you will probably need to find additional data at some point.

A cost-benefit analysis often must compare various scenarios and determine the relative advantage of each. We will provide an example of this kind of analysis based on Robert Herendeen's *Ecological Numeracy* with the values changed to apply to Connecticut in 2007.[1] The goal is to compare the total cost of providing equal amounts of residential lighting using either incandescent or compact fluorescent bulbs (CFLs).

The incandescent bulbs are less expensive to buy initially but require much more electricity to operate. Incandescent bulbs also have a shorter life expectancy of 800 hours versus 12,000 hours for the CFLs. We will examine the cost of each option over the thirty-year life of the power plant using enough bulbs to maintain a constant use rate of 1 kilowatt (kW).

Consider the total cost of operating thirteen 75 W incandescent bulbs for a year. This requires a continuous supply of 1,000 kW of power and includes the cost of both the electricity and the cost of the bulbs. There are 8,760 hours in a year, so 8,760 kW hours of electricity are used. Because the average life of the bulbs is only 800 hours, each bulb will need to be replaced about eleven times. That means a total of 146 bulbs will be used. Over a thirty-year period, you will need 4,369 bulbs, which cost $0.75 each, for a total cost of $3,276 for the bulbs.

The use of CFLs to produce the same total light requires thirteen 17W bulbs using only 0.227 kW of power. Over the course of one year, they require 1,988 kW hours of power. Each CFL costs $12 but has an average lifetime of 12,000 hours. During the first year, no bulbs need to be replaced; over the course of thirty years, only 292 bulbs are used. The cost of the CFL bulbs is then $3,507. The consumer pays about $10 per year more for CFL bulbs compared with incandescent bulbs.

The savings to the consumer are in the cost of electricity. Assuming a cost of $0.17 per kilowatt hour, the incandescent bulbs will cost $1,489 to operate; the CFLs will cost $338. The consumer saves a bit more than $1,000 per year by using the thirteen CFLs instead of incandescent bulbs. The utility company also saves money because it only needs to build a plant to supply a quarter of the electricity that is required for the incandescent bulbs. If it costs $6,600 per kilowatt hour to build a power plant, then the company saves about $5,100 in construction costs. Spread over the thirty-year lifetime that is $170 per year less that the utility must spend. These calculations are summarized in Table 19.

The case for using CFLs seems compelling. The total cost over thirty years is about one-fourth that of the incandescent bulbs. In addition, the total emission of acid pollutants, carbon dioxide, nuclear waste, or whatever else negative occurs due to the production of the power will be one-fourth as great with the CFLs. The question then is why are people still buying incandescent bulbs?

There are numerous answers to the question. Part of the reason is inertia. People buy what they are used to buying. And some people claim to like the light from the incandescent bulbs better. But

---

1. R. A. Herendeen, *Ecological Numeracy: Quantitative Analysis of Environmental Issues* (New York: John Wiley & Sons, 1998), chap. 5.

TABLE 19 Cost comparison for electric lights over thirty years

Parameter	Incandescent	Fluorescent
Capital cost of power plant construction (30-year life) included in cost of electricity that must be included in cost of electricity	$6,600	$1,500
Purchase of bulbs	13 bulbs × 11 bulbs/year × $0.75/bulb × 30 years = $3,276*	13 bulbs × 0.733 bulbs/year × $12/bulb × 30 years = $3,507
Cost of electricity ($0.17/kWh)	$44,676	$10,141
Total lifetime cost to consumer	$47,962	$13,648

Note: kWh, kilowatt hours.
* Full equation: 13.3 bulbs × 8,760 hours/year × 30 years × 1 bulb/800 hours × $0.75/bulb.

there is a deeper issue that also needs to be considered. There are differences in when the money is spent. Money that is not spent can be invested and will earn interest or can be spent for other things. For the power company, the incandescent bulbs require a big upfront investment in the power plant. They would prefer not to spend that and may spend money to encourage customers to buy CFLs. The consumer, on the other hand, sees the initial $12 price tag for the CFL versus 75 cents for the incandescent. The latter leaves $11.25 in her pocket until the bulb burns out.

In fact, the customer does not ever save money on the purchase price of the light bulbs. It is only when the electric bill comes in that the savings accrue. If one partner buys the bulbs and the other partner pays the electric bill, they may never make the connection and will waste $34,000.

To see the effect of time on the costs and benefits, it is instructive to examine a single bulb that will be operated for 10,000 hours over the course of five years. The total capital cost to the consumer occurs in the first year when she buys the bulb for $12. This will be the only bulb she buys in those five years. To have used incandescent bulbs would have cost only $1.88 each year for a five-year total of $9.37 (10,000 hours × 1 bulb/800 hours = 12.5 bulbs used × $0.75 = $9.37). The energy savings are dependent on the price of electricity. Using 17 cents/kW hour, the savings on each year's electric bill are $19.72 (75 − 17) watts × 1 kW/1,000 watts × 2,000 hours/year × $0.17/kW hour). This means she actually saves $9.60 in the first year and $21.60 each year for the next four years by having bought the CFL.

When economists look at this type of analysis, they add several additional factors. They include the time value of the money spent on the initial capital outlay and the fact that money not spent could have been invested. They also adjust the value of the money spent and saved for the predicted effects of inflation on the value of the money. In making arguments for costs and benefits, two economists can reach different conclusions based on how they estimate the discounting of the future value of money spent and saved.

Appendix 4. Study Questions for Reading Assignments

QUESTIONS ON ENVIRONMENTAL PHILOSOPHY

1. Compare Aristotle and Plato to the book of Genesis' position on man's relationship to the world.
2. What three alternatives does Marsh suggest for the relationship of people to Nature? How do these correspond to the 1951 UN statement on human progress?
3. Compare Naess's concept of "Milieu" to Leopold's idea of community.
4. What is the function of an ethic?
5. How does an ethic differ from an instinct? What about modern life requires the development of an environmental ethic?
6. What is the basic premise of an ethic?
7. What is the community on which Leopold's Land Ethic is based?
8. Why cannot humans be "conquerors" of nature?
9. Why are economic arguments insufficient to develop an environmental ethic?
10. How does an ethic function to modify behavior?
11. Marsh and Leopold, almost a century apart, commented on the effects of human habitation on various areas. Which areas do they find have the most resilient biological communities and which have the least?
12. What is the number one factor that determines the damage of human impact on the ecosystem?
13. How does an ethic develop and operate? (This has direct relevance to the way in which the Swedish and Norwegian factions should approach the problem of control of acid rain.)
14. Compare the traditional worldview with that of Deep Ecology. What are the potential weaknesses of each?

QUESTIONS ON LOVELOCK'S GAIA HYPOTHESIS

Chapter 1. How did the search for life on Mars lead to the Gaia hypothesis?

Chapter 2. How does the relative constancy of the earth's temperature support Lovelock's argument for a global organism in control of the environment?

Chapter 3. How does the comparison of the atmospheres of Mars, Earth, and Venus support the Gaia hypothesis?

Chapter 4. What is the difference between positive and negative feedback loops in a cybernetic system? What examples of feedback systems does Lovelock use?

Chapter 5. Volcanos and some biological processes produce sulfate and sulfuric acid. Nitrates and nitric acid are also produced by oxidation of nitrogen. What does Gaia do to neutralize this acid and prevent the Earth from becoming too acidic for life? How does the anthropogenic production of sulfur and nitrogen oxides from fossil fuel combustion threaten the balance of pH of the biosphere?

Chapter 6. This chapter begins with a question: "Why is the sea salty?" Why is this the wrong question to ask?

Chapter 7. Lovelock conducted the early studies that showed chlorofluorocarbons to be spread around the globe and supported the efforts to reduce human emissions of these freon gases. What is his opinion of this effort and how does he rationalize this in terms of Gaia? What are the most serious human pollution challenges according to Lovelock? What is the role of information technology/communication in humanity's role in Gaia? What is the difference between "spaceship earth" and Gaia?

Climate Change in Copenhagen, December 2009

DAVID E. HENDERSON

Contents

1. HISTORICAL BACKGROUND / 129

    Overview / 129

    Prologue / 130

    Time Line / 130

    Historical Context / 131

        The Kyoto Mechanisms / 131

    Scientific Background / 132

        Solar Energy Input / 133

        Moving Heat Around: Convection in the Atmosphere / 134

        Moving Heat in the Oceans: Thermohaline Circulation / 136

        Greenhouse Effect / 136

            How It Works / 136

            Greenhouse Gases / 136

        The Carbon Cycle: Sources and Sinks / 139

        Keeping the Temperature Constant: Homeostasis and Feedback/Forcing / 141

        Sources of Greenhouse Gases / 143

        Conclusion / 148

        What Can Be Done about Greenhouse Gases? / 149

        Goals for Minimizing Climate Change / 149

    Political Background: International Environmental Treaties / 150

        Pollution Reduction Mechanisms / 150

            Command and Control / 151

            Polluter Pays / 151

            Cap and Trade / 151

        Compliance / 152

    Glossary and Guide to Abbreviations / 152

2. THE GAME / 153

    Major Issues for Debate / 153

        Issue 1: Limits on Greenhouse Gases / 153

        Issue 2: Nature of the Agreement / 155

Issue 3: Transparency / 155
Issue 4: Money Transfer from Developed to Developing Countries / 156
Issue 5: Money to Protect Forests / 156
Issue 6: Remediation / 156
Rules and Procedures / 156
Conference Protocol / 156
Private Meetings / 156
Voting at the Conference / 156
After the Conference: National Approval / 157
Outline of the Game / 157
Game Session 1: Opening Session / 157
Game Sessions 2 (and 3): General Debate and Approval of a Treaty / 157
Before the Final Session / 157
Debriefing / 157
Assignments / 157

3. ROLES AND FACTIONS / 159

Indeterminate Countries / 159
Greens: Countries That Want the Strongest Possible Treaty / 160
Opponents of Any Treaty / 161

4. CORE TEXTS AND SUPPLEMENTAL READINGS / 162

Supplemental Sources / 162

Acknowledgments / 164
Appendix 1. Greenhouse Gases / 165
   Absorption of IR Radiation / 166
Appendix 2. Chemicals in Fossil Fuels / 168
   Larger Molecules of Carbon, Oxygen, and Hydrogen / 168
      Naming Hydrocarbons / 169
   Organic Molecules Containing Oxygen / 170
Appendix 3. Quantitative Look at Combustion Reactions / 172
   Stoichiometry Calculations and Weights of Reactants and Products / 173

    Counting by Weighing / 173
    Applying the Mole Concept / 174
  Energy Associated with Combustion Reactions / 176
    Calorimetry / 176
  Calculations Using Bond Energies / 176
Appendix 4. Leaked Draft Document by Danish Delegates / 180

Figures and Tables

FIGURES

1. Solar Energy Inputs and Outputs from Earth / 134

2. Atmospheric Circulation Cells / 135

3. Thermohaline Circulation / 137

4. Infrared Spectrum of Earth as Seen from Space / 138

5. The Carbon Cycle / 140

6. Homeostasis of Body Temperature—Keeping the Temperature Constant / 141

7. Negative Feedback System in Climate / 142

8. Ice Positive Feedback System / 143

9. Permafrost Positive Feedback / 144

10. Sources of Carbon Dioxide / 145

11. Traditional Electrical Generation Process / 146

12. Cogeneration: Combined Heat and Power Generator / 147

13. Well to Wheels Efficiency of Cars / 147

14. Sources of Methane / 148

15. Estimated Relationship between Final Equilibrium Carbon Dioxide Levels and Global Temperature / 154

16. Percentage Reduction in Global Greenhouse Gas Emissions Required to Reach Various Equilibrium Concentrations / 155

17. The Electromagnetic Spectrum / 166

18. Vibrational Modes for Carbon Dioxide / 166

19. Vibrational Modes for Water / 167

20. Example of Hydrocarbon Structures for Naming / 170

21. Structure of the Alcohol Group / 170

22. Possible Structures of a Four-Carbon Molecule / 171

23. Bomb Calorimeter / 177

TABLES

1. Greenhouse Gases and Their Relative Effectiveness / 139

2. Treaty Characteristics for Debate / 154

3. Names of Normal Hydrocarbons / 169

4. Energies Associated with Specific Chemical Bonds / 178

1
Historical Background

OVERVIEW

This game will take you to Copenhagen where you will participate in the negotiations to formulate a global climate treaty. These negotiations have been going on for years but have failed to produce a treaty that the majority of polluters are willing to sign. There is great optimism that Copenhagen will be different. As a result, many heads of state have chosen to attend the final few days of the meeting, either to increase the chance of success or to ensure their national interests are properly represented.

For most students in this game, your role is not that of a scientist but a politician. The educational training of your character may be similar to your current academic major. Your character may have studied economics, political science, history, or even art. There are a number of questions that all conference participants need to understand. Without a solid understanding of these, you may be misled by things you read or the arguments you hear.

1. How does the climate system work to keep a relatively constant climate on Earth?
2. What is the greenhouse effect, and how does it affect climate?
3. What are greenhouse gases, and what are the properties of molecules that make them greenhouse gases?
4. What chemicals and chemical reactions are used to power our society, and how do they lead to the production of greenhouse gases?
5. Which fuels produce the least greenhouse gases?
6. What strategies are available to reduce the use of fossil fuel?
7. What are the potential risks and benefits of a warmer climate?
8. What actions have nations already taken to deal with the climate change?
9. Which nations are most responsible for greenhouse gases?

This game is designed to take you outside yourself and allow you to place all the arguments on the table for examination. Players will eventually decide which arguments to accept based on the quality of the evidence presented. Some players need to argue as climate change skeptics; they should do so with passion and enthusiasm. Whatever your role, you must play it well and find the best possible evidence to support your arguments.

PROLOGUE

Yesterday, you arrived in Copenhagen after a long day of travel. You were interested to see the extent to which Denmark has become serious about climate change. The wind farm at the harbor was visible out the window of your airplane as your plane circled for landing. You don't remember ever seeing so many windmills in one place in your life. You were also struck by the row of charging stations for electric cars outside the hotel and by the dense bicycle traffic on the streets as commuters headed home from work. You have traveled the world, and the irony is clear. In Denmark, people are abandoning their cars for bicycles, and in the People's Republic of China and India people are parking their bicycles and buying cars. You have also seen farms in the United States where an old Depression era windmill stands rusting on one side of the barn while the farmer now lives off the rental he gets from the dozens of windmills in his pastures.

As a veteran of climate change talks, you come to Copenhagen with a feeling of anticipation. The Copenhagen Climate Conference promises to be different that those that went before it. There is a new administration in the United States, and President Barack Obama campaigned on doing something about climate change. The United States has been the major sticking point in all previous meetings. They won't do anything before China and India, and China and India won't do anything before the United States. This has made the climate deniers ecstatic and has frustrated the environmentalists to no end. Possibly this time will be different. You need to be prepared to meet the expectations of your government, which means preparing your arguments carefully. You will need to check for the latest data and try to prepare graphs that show your government's position in the best light while making your adversaries look bad.

Before you get down to the serious work of making your PowerPoint slides for your talk, you should probably plan your strategy with the rest of your faction. You pick up the hotel phone and make a few calls. After a shower and a change of clothes, you are off to meet your faction at Tivoli. It should be a lovely place to plan arguments and divide the work.

Arriving at Tivoli, you walk past the brightly lit buildings and enjoy the reflections of lights in the lake. Finally you make your way past the Pagoda to the small out-of-the-way restaurant and see your friends at a table. It is quite cold, so you are happy to get inside. They have already ordered a round of Aquavit and have a carafe of wine as well. You allow the liquor to warm you and begin to plot your strategy for the conference.

TIME LINE

The first meeting to produce a climate treaty was held in 1995 in Berlin and is known as the Conference of Parties (COP1). The 1997 conference in Kyoto, Japan, was COP3 and led to a treaty that was adopted and ratified by 192 countries. The Kyoto Protocol entered into force on February 16, 2005. The detailed rules for the implementation of the Protocol were adopted at COP7 in Marrakesh in 2001, and are called the Marrakesh Accords.

However, the Kyoto Protocol was never approved by the largest polluter, the United States, due to opposition from the U.S. Congress. Even China and India signed the protocol, which required little of them. Therefore, it had only a limited effect on greenhouse pollutants, and climate change continues to accelerate. Even the nations that did sign it have had difficulty meeting the obligations of the treaty, and some have withdrawn. Furthermore, the Kyoto treaty was only an interim treaty and expires in 2012.

The setting for this game is the Copenhagen Climate Conference (COP15) held in December 2009. This conference was the culmination of many smaller working group meetings designed to produce a treaty that could be presented to the entire world. This is actually the fifteen meeting of one track of these negotiations and the fifth of another.

HISTORICAL CONTEXT

The first international treaty to deal with air pollution was the Montreal Protocol. This addressed the damage to the stratospheric ozone layer caused by chlorofluorocarbons (CFCs). These chemicals were used in most refrigerators and air conditioners. They had also become important for making plastic foam and were used as propellants for aerosol cans of all types and as solvents for cleaning metal and electronic parts. Research by Rowland and Molina at the University of California was the first hint that these widely used chemicals could destroy the layer of ozone in the upper atmosphere. Their research was honored with the 1995 Nobel Prize in Chemistry.[1] This ozone absorbs ultraviolet rays from the sun that can cause skin cancer in people and damage plant leaves. Rowland and Molina called for a ban on the use of CFCs in 1974.[2] Over the next decade, more scientific studies confirmed the danger of CFCs, but industry groups refused to accept the need to eliminate them. Finally, in 1984 a group of British researchers discovered a large hole in the ozone layer over Antarctica. This led to a 1985 meeting in Vienna that provided a framework for reducing CFC use. The discovery of the ozone hole galvanized public opinion, and by 1987 the Montreal Protocol was signed. The treaty phased out the production and use of CFCs and provided a period for the use of hydrochlorofluorocarbons (HCFCs) and alternative groups of chemicals that do not attack the ozone layer. The treaty also provided a timetable for replacing HCFC.

Following on the heels of the success of the Montreal Protocol, an international effort to address climate change began in 1988 with the formation of the Intergovernmental Panel on Climate Change (IPCC). The IPCC issued a report in 1990 calling for a global treaty to deal with the threat of climate change, and international negotiations began in 1991. This group prepared the documents that were then discussed and approved at an international conference in Rio de Janeiro, Brazil, in 1992. This meeting was held with a sense of great optimism, and the UN Framework Convention on Climate Change (UNFCCC) and several other conventions on climate were submitted to the nations of the world for their approval. The UNFCCC took effect in 1994 and instituted an ongoing process of international meetings to address climate change. The goal was to produce a comprehensive international treaty similar to the Montreal Protocol to reduce emissions of gases responsible for climate change.

The Kyoto Mechanisms

Under the Kyoto Treaty, each country must meet their targets primarily through internal reductions in emissions.[3] However, the Kyoto Protocol offered three additional ways that countries can meet their targeted reductions.

---

1. Nobel Prize, "The Nobel Prize in Chemistry 1995," press release, October 11, 1995, https://www.nobelprize.org/nobel_prizes/chemistry/laureates/1995/press.html.

2. M. J. Molina and F. S. Rowland, "Stratospheric Sink for Chlorofluoromethanes: Chlorine Atom-Catalysed Destruction of Ozone," *Nature* 249, no. 5460 (1974): 810–12.

3. United Nations Framework Convention on Climate Change, "Kyoto Protocol," http://unfccc.int/kyoto_protocol/items/2830.php.

- *Emissions trading.* Known as "the carbon market," emissions trading allows a country to reduce its emissions more than necessary and sell emission credits to a country that finds it more expensive to reduce emissions. This is thought to allow the most cost-effective reductions because it encourages countries to make the least expensive reductions as fast as possible to sell carbon credits.
- *Clean development mechanism (CDM).* This mechanism allows a country to pay for emission reductions in a developing country instead of expensive ones in their own country. They receive credits that allow them to continue to emit greenhouse gases equivalent to what they reduce in the developing country.
- *Joint implementation (JI).* Joint implementation is similar to CDM, but it includes projects such as planting forests in one country to capture carbon to be paid for by another country to count toward their pollution reduction.

The mechanisms are intended to help stimulate green investment and help parties meet their emission targets in a cost-effective way. In order for these mechanisms to work, there must be careful monitoring of emissions, The Kyoto Protocol required that *actual emissions had to be monitored* and precise records had to be kept of the trades carried out. This is done using a variety of methods.

- The UN Climate Change Secretariat, based in Bonn, Germany, keeps an international transaction log to verify that emission trades are consistent with the rules of the Protocol.
- The parties to the treaty submit annual emission inventories and national reports.
- A compliance system ensures that parties are meeting their commitments and helps them to meet their commitments if they have problems doing so.

Finally, the Kyoto Protocol is also designed to *assist countries in adapting* to the adverse effects of climate change. The Adaptation Fund was established to finance adaptation projects and programs in developing countries that are parties to the Kyoto Protocol. The fund is financed mainly with a share of proceeds from CDM project activities.

Emissions targets for the Kyoto protocol varied from a reduction of 8 percent relative to the 1990 benchmark by the European Union countries and others, to an increase of 10 percent relative to 1990 by Iceland and an increase of 8 percent by Australia. These targets were set by the individual governments rather than imposed by the agreement. There was no overall structure to ensure that these targets led to the desired outcome for the climate.

The Copenhagen meeting is COP15 and needs to plot a course forward after Kyoto because the Kyoto Protocol expires in 2012. To be successful, COP15 must produce a treaty that can be accepted by the United States, China, and India. The Copenhagen meeting opens with a real optimism that a final treaty is at hand. This is where the next treaty will succeed or fail. The stakes are high enough that the leaders of over 100 nations are expected to attend, and there are rumors that even President Obama may attend if his voice is needed to close the deal.

SCIENTIFIC BACKGROUND

There are a number of technical aspects of climate that are not disputed. This section will provide a brief introduction to this material in an effort to allow you to understand the fundamentals involved in the earth's climate system. These topics include the energy balance of the earth, the way heat is moved around, how greenhouse gases make the earth habitable, and the way carbon cycles in the ecosphere.

Solar Energy Input

Virtually all the heat available on Earth comes from the sun. The only other source is geothermal energy moving up from the hot center of the Earth, which is residual heat from the initial molten earth and heat from radioactive decay. The sun has gradually become warmer over geological time scales, but within human history it has remained relatively constant.

There are variations due to "weather" on the sun itself in the form of sunspots and solar flares. The primary cycle for these variations is the eleven-year sunspot cycle. When there are many sunspots, the sun provides slightly more heat to the Earth than when there are no sunspots. This oscillates through an eleven-year cycle, though sometimes the cycle shifts or misses a beat, as it did in 2011 when expected sunspots failed to appear on schedule. There is also a cycle of about 110,000 years due to a periodic variation in the orbit of the earth around the sun.[4]

There are two ways to look at the energy coming from the sun: as a percentage of the total or in absolute terms using heat units. We will use watts (W) for our heat unit. You are probably familiar with wattage of light bulbs: the small bulbs used as nightlights are 7 W, and typical lamps range from 60 W to 100 W. On this scale, the sun provides about 342 W of total energy in the form of heat and light to every square meter of the earth's surface.

Equally important is what happens to the sun's energy when it reaches the Earth. Figure 1 below shows the major processes that occur and the wattage for each.[5]

The major factors affecting the sun's energy input to Earth are the amounts reflected and the amounts absorbed. Reflected radiation goes directly back into space. The earth has a relatively high albedo of 30 to 35 percent. (*Albedo* is the ratio of solar energy reflected over total incident energy.) Clouds and particles in the atmosphere reflect about 31 percent of the total. This leaves 69 percent to be absorbed by the atmosphere and the surface. This absorbed energy is converted to heat. Most of the absorbed light heats the surface of the Earth, and the surface in turn heats the bottom of the atmosphere. The warm air is lighter and rises in thermals, carrying some heat up through the lower atmosphere. The 235 W/m^2 that is absorbed by the surface and the atmosphere (168 + 67) provides heat energy that drives the weather and keeps the Earth comfortable for life.

The second factor in the energy balance is the energy emitted by the Earth back into space. All objects emit electromagnetic radiation (heat and light) in a way that depends on their temperature. The exact nature of the spectrum of this radiation has been known since the end of the nineteenth century and is described as *black body radiation*. The sun, with an effective temperature of 5,700 K,[6] gives off mostly visible and ultraviolet light. Your body at 310 K gives off infrared radiation that is detected by night vision goggles and thermal imaging systems. The global average surface temperature is about 288 K (15°C or 59°F), and the Earth gives off infrared (IR) light (the form of electromagnetic radiation we perceive as heat) consistent with this temperature.

Figure 1 shows that the total energy in the form of IR radiation leaving the atmosphere is 235 W/m^2 (165 + 30 + 40). When you add the 107 W/m^2 that is directly reflected, you get 342 W/m^2, which is the same as the total incoming solar radiation. The incoming and outgoing radiation must be the same, or the earth's temperature would increase or decrease. If the outgoing radiation is greater than the incoming radiation, as it is at night, the earth

---

4. Milankovitch cycles.

5. J. T. Kiehl and K. E. Trenberth, "Earth's Annual Global Mean Energy Budget," *Bulletin of the American Meteorological Society* 78, no. 2 (1997): 197–208.

6. Kelvin is a measure of temperature that eliminates negative temperatures. Zero K is absolute zero, the lowest possible temperature. To convert from Celsius to Kelvin, add 273. That means that room temperature of around 22°C is equivalent to 295 K.

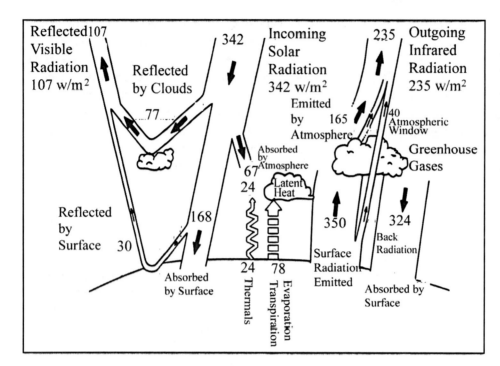

FIGURE 1 Solar energy inputs and outputs from Earth. (Adapted from Kiehl, J. T., and Trenberth, K. E. "Earth's Annual Global Mean Energy Budget," *Bulletin of the American Meteorological Society* 78, no. 2 [1997]: 197–208.)

gets cooler. If the incoming radiation is greater than the outgoing radiation, the earth gets warmer. So the change in temperature from day to night reflects this balance between incoming and outgoing radiation.

Greenhouse gases are an important part of this overall process. Again, refer to Figure 1 and you will see that greenhouse gases absorb a very large amount of energy being emitted by the Earth. The Earth's surface actually emits 390 W/m² but greenhouse gases trap 350 W of this. Greenhouse gases are molecules that absorb IR radiation. Each molecule has specific wavelengths absorbed and transmitted. Most of this absorbed radiation warms the greenhouse molecules and is reemitted back down to the surface (325 W/m²). Only a part of the surface emissions escapes directly to space (40 W/m²). Most are absorbed in the atmosphere, and some of that goes from the atmosphere out to space (165 + 30) and some comes back to the surface (324). Appendix 1 has more details about what makes some molecules greenhouse gases.

The final process that moves significant heat is the evaporation of water from the surface. Evaporating water cools the Earth just as perspiration evaporating from your skin cools your body. The rising water vapor carries this heat into the atmosphere. The 78 W/m² labeled evaporation/transpiration is the heat moved into the atmosphere by evaporation of water from the surface and transpiration of water out of the leaves of plants. This heat has a major impact on weather. When the water vapor in the atmosphere condenses to form clouds, the heat is released into the atmosphere. The major driver of storms is the energy from this process.

Moving Heat Around: Convection in the Atmosphere

The region around the Equator is more directly exposed to the sun than any other. The polar regions receive the least light due to the glancing angle of

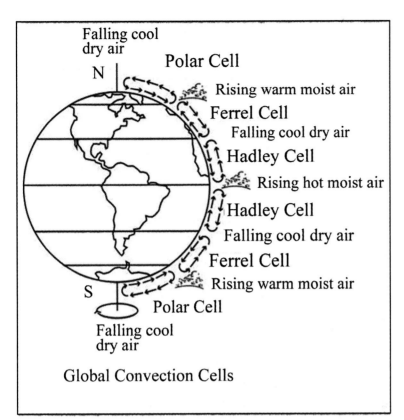

FIGURE 2 Atmospheric circulation cells.

the sun. Thus, more heat is received at the Equator than at the poles. The rising air warmed by the surface (thermals, 24 W/m² in Figure 1) and the heat transport from evaporation (78 W/m²) moves heat into the atmosphere and is, therefore, greatest in the tropics near the Equator. The imbalance in temperature between different regions drives wind and weather. There are specific patterns of rising and falling air shown in Figure 2 that have remained in the same places throughout human history.

Warm, moist air at the Equator rises and moves north and south. The rotation of the Earth causes the actual winds to blow at an angle rather than directly north and south. By the time the air reaches about 30 degrees north and south latitude, the air has cooled and dried and sinks back down. The region from the Equator to the point where the air sinks back down is called the Tropics. The Tropics are wet due to the rainfall that dries the air moving north and south. This leads the areas around 30 degrees to be dryer and cooler due to the falling air. The cooler air moves back to the Equator, forming a circulation called the Hadley Cell. The winds in the middle of the Hadley Cell from 10 to 20 degrees latitude are called the *trade winds* and blow east to west, and the direction of these winds was used by sailors.

The second convection cell begins at 30 degrees and goes up to 60 degrees latitude. This air moves north in the Northern Hemisphere and picks up heat and moisture from the surface. At 60 degrees it rises and then moves back south as it loses its heat and moisture. This region is known as the Temperate Zone and is less hot and moist than the Tropics. This circulation is named the Ferrel Cell, and the surface winds in this region blow west to east. Sailors know this as the Westerly Winds and have used these winds to move back and forth across the oceans since Columbus. Sailors in the Northern Hemisphere would go to the tropics to catch the Trade Winds to go west to America, and then would go north to the temperate zone to catch the Westerlies to go east to Europe.

The regions above 60 north and below 60 south latitude are called the Arctic and Antarctic, or the Polar Regions. The third convection cell lies at the poles and forms the Polar Cell. At both poles, the winds circulate around the globe in the Polar Vortex. The Polar Vortex tends to keep cold air at the poles and prevent it from moving toward the Equator.

The weather and the climate of various regions depend primarily on where they lie in these convection cells. Regions of desert and high moisture are all dependent on this overall circulation pattern. Weather is the short-term transport of heat and moisture in the form of rain, wind, and sunshine. Climate reflects the longer time-scale variations. But the climate of any particular region depends on how much solar energy it receives due to the angle of the sun and where the region lies in the circulation cells.

Moving Heat in the Oceans: Thermohaline Circulation

The other circulating fluid on Earth is the water in the oceans. Oceans have their own climate and weather, with long-term and short-term changes, respectively. The amount of moisture in the ocean cannot change, but the amount of salt dissolved in the water does change. And the combination of salinity (saltiness) and temperature drives a pattern of circulating water that parallels the convection cells in the atmosphere. Colder and saltier water is heavy and sinks to the bottom while warmer and less salty water rises.

The pattern of water circulation shown in Figure 3 shows how shallow, warm currents carry heat from the tropics to the poles and how the water returns in deep, cold currents. The most familiar warm current is the Gulf Stream, which is the shaded arrow on the upper left of the figure flowing northward along the Americas and ending in northern Europe. It is estimated that as much as 30 percent of the heat in the atmosphere in northern Europe—including Great Britain, Ireland, and Norway—comes from the Gulf Stream. Great Britain lies much farther north than other areas with similar climate, and it owes its moderate climate to the Gulf Stream. For example, Edinburg in Scotland is at the same latitude as Anchorage, Alaska, but it has a much different climate.

Just as the regional weather varies from year to year, so do these ocean currents. The combined effects of the convection cells of the atmosphere and the ocean currents account for the climate of the various regions of the earth.

Greenhouse Effect

How It Works

The greenhouse effect is the name for the process we have already described by which certain gases in the atmosphere absorb IR radiation. As shown in Figure 1, some of this radiation is re-radiated back to the surface, and some goes out into space. Greenhouse gases are a critical factor in making the Earth habitable. It has been calculated that without greenhouse gases, the surface temperature would be 30°C lower, shifting from about 15°C to −15°C. The absence of greenhouse gases would allow more heat to escape to space and reduce the temperature so much that there would be little or no liquid water, only ice. In fact, some geological data suggest that at one point in the distant past, there was ice over virtually the entire surface of the Earth. So we have to appreciate the presence of these gases for keeping things comfortable for us.

Greenhouse Gases

A number of gases act as greenhouse gases. More details on what makes a gas a greenhouse gas are provided in Appendix 1. Carbon dioxide (CO_2) is the one that is most prominent in the discussion

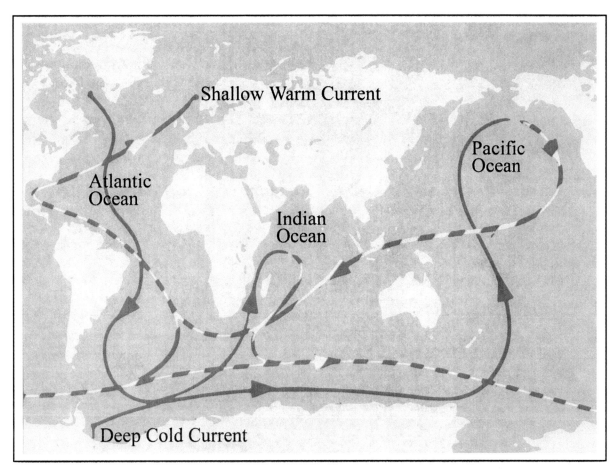

FIGURE 3 Thermohaline circulation. Heat is released to the atmosphere in the Arctic and Antarctic regions. Solid lines are cold currents.

of climate change. But methane (CH_4, natural gas) and water vapor also contribute to the Greenhouse Effect. What these gases have in common is that they absorb IR radiation. Nitrogen and oxygen, the two primary gases in the atmosphere, are not greenhouse gases. In order to absorb IR radiation, a molecule needs to have at least two different kinds of atoms or at least three atoms.

We can explore the effect of the absorption of heat in the form of IR radiation by looking at Figure 4, the IR spectrum of Earth seen from space. This is a complicated figure with many lines. We will approach it one step at a time.

First, look at the five solid curves in the figure labeled with temperatures from 175 to 300 K. (This is a range from about −100° to 27°C.) These show what is called the black body emission spectra of objects at various temperatures. Every solid object emits light with a spectrum that depends on its temperature. You can think of this as what happens when you turn up the dimmer on an incandescent light bulb. The light gets brighter and whiter as you apply more power. As the temperature of the filament increases, the emission becomes stronger (brighter, meaning more total emission) and the wave number of maximum emission shifts to larger values (whiter). Looking only at the solid, smooth curves for temperatures from 175–300 K, you see this effect. The temperature of objects can be measured by observing their

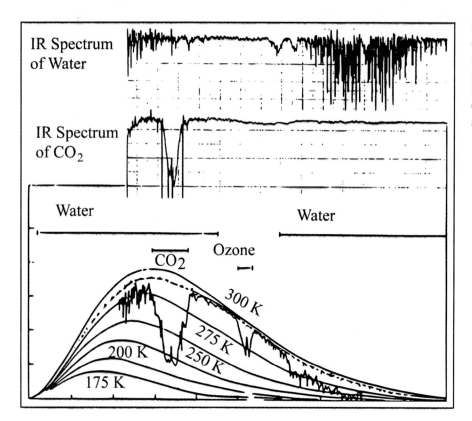

FIGURE 4 Infrared spectrum of Earth as seen from space. (Adapted from Hansel et al. [NASA], "IRIS/Nimbus-4 Level 1 Radiance Data V001." Overlay of IR spectra by author.)

IR spectra. This is how the fever thermometers placed in the ear work.

Now look at the dotted line between the smooth curves for 275 and 300 K. This line shows what the emission spectrum of the Earth would look like in the absence of any greenhouse gases.[7] The earth's current temperature averages around 285 K, so this line lies between 275 and 300 K.

Finally, examine the jagged line below the dotted line. This is the actual IR emission spectrum of the Earth measured by a satellite at night. When we compare the actual emission spectrum, the jagged line, to the dotted line, it is clear that there are some regions of the spectrum where the intensity is lower. These parts of the spectrum are being absorbed, resulting in fewer emissions reaching space. The reduced emissions mean the Earth cannot emit as much energy, and the result is that the Earth is warmer due to the heat trapped by the greenhouse gases. The energy absorbed by these dips in the emission spectrum is enough to raise the global temperature about 30 degrees.

Now we need to identify which gases are responsible for the absorption of the light leaving the earth. The largest absorption peak matches the IR absorption of CO_2 that is shown in the inserted spectrum at the top of the figure. Note that this peak is very close to the place where the Earth's emissions reach a maximum. As a result CO_2 is the most effective greenhouse gas. Its absorption peak occurs in the part of the spectrum where the earth

7. R. A. Hansel et al., "IRIS/Nimbus-4 Level 1 Radiance Data V001," (Greenbelt, Md.: Goddard Earth Sciences Data and Information Services Center [GES DISC]), https://disc.gsfc.nasa.gov/datacollection/IRISN4RAD_001.html.

has the greatest IR emission. The heat that is absorbed by the CO_2 means that the Earth must get warmer to bring the incoming solar energy and the outgoing IR radiation from the earth (black body radiation) into balance.

The absorption spectrum of water is also shown in the insert at the top. This spectrum has a large number of very sharp peaks. The strongest absorption of water vapor occurs where the earth has less emission than in the region where CO_2 absorbs. The water absorptions account for the many sharp irregularities in the earth emission spectrum in the two regions where water absorbs.

There is also a peak due to ozone (O_3). Ozone absorbs outgoing IR in both the lower atmosphere where ozone is a significant pollutant and in the stratosphere where it also absorbs the dangerous ultraviolet light from the sun.

The relative importance of the greenhouse gases is shown in Table 1 along with the approximate concentrations at the time of the Copenhagen meeting. Of the five gases in the table, CO_2 appears to have the lowest relative effectiveness as a greenhouse gas. Nevertheless, its much higher concentration means that the effect of CO_2 on climate is greater than all the other gases combined. As a result, this gas receives the most attention. The total impact column multiplies the concentration by the relative effectiveness. This also makes clear that changes in CO_2 have the greatest potential to cause or remedy climate problems.

Water is not included in Table 1. The water vapor in the atmosphere is directly linked to temperature and is not related to pollution. Also, the value is self-limiting through a variety of natural processes. However, you can experience the impact of water vapor by observing the nighttime temperatures on clear dry nights and on humid and cloudy nights. The former are very cold, and the latter are much warmer.

The Carbon Cycle: Sources and Sinks

Carbon dioxide is part of the global carbon cycle. Considerable scientific study has been done to try to account for all of the carbon on the earth and the ways it moves through the earth systems.

Figure 5 shows the current knowledge about the amount of carbon in various places on earth and the rate at which it moves from one area to another. The units are gigatons of carbon with 1 Gt=1 billion tons. Fossil fuels and carbonate minerals (like limestone) hold the greatest amounts of carbon. Fossil fuels account for 3,700 Gt. Carbonate minerals are not shown but represent a similar amount.

TABLE 1 Greenhouse gases and their relative effectiveness

Gas	Preindustrial Concentration (ppm)	Present Concentration (ppm)	Relative Effectiveness $CO_2=1$	Total Impact: Concentration Times Effectiveness
CO_2	280	390	1	390
CH_4	0.7	1.76	25	44
N_2O	0.275	0.32	298	95
CFCs	0	0.00056	8,100	4.5
SF_6	0	0.000006	22,800	0.14

Note: CFCs = chlorofluorocarbons; CH_4 = methane; CO_2 = carbon dioxide; N_2O = nitrous oxide; SF_6 = sulfur hexafluoride.

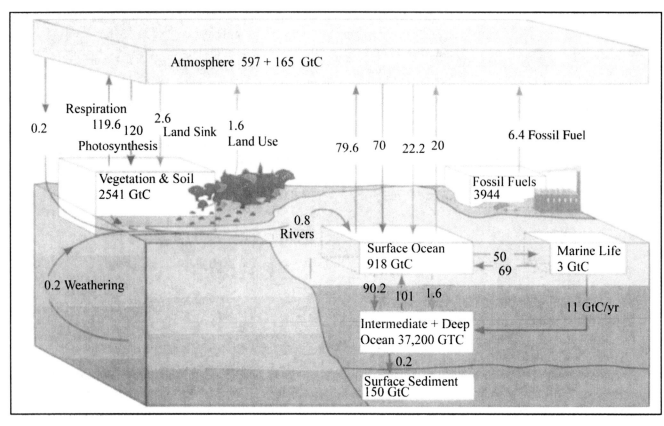

FIGURE 5 The carbon cycle. The arrows indicate flows of carbon in gigatons per year.

The deep oceans are also major repositories of carbon at 37,000 Gt. Compared with these reservoirs of carbon, the amount in circulation is relatively small. The atmosphere contains only about 600 Gt, and the surface water has 900 Gt.

The sources of carbon are physical or biological processes that move carbon into the air. Sinks of carbon are physical and biological processes that remove carbon from the atmosphere and tie it up in plant tissue, the oceans, and rocks. Respiration is a source, and photosynthesis (gross primary production) is a sink. Respiration is the process by which animals and to a lesser extent plants burn carbohydrates for energy and release CO_2. Photosynthesis is the process of plants turning CO_2, water, and sunshine into sugar, starch, and cellulose in plant tissue. These two opposite processes are in close balance, with slightly more photosynthesis than respiration in the figure. If one plants more trees, then photosynthesis increases and more carbon is removed than is created by respiration. If trees are cut down to raise cattle, then respiration may exceed photosynthesis. Cutting trees for deforestation has two effects on the carbon cycle. It reduces photosynthesis, and the cut trees are either burned or rot. Both processes return the carbon in the trees to the atmosphere as CO_2. Currently, deforestation exceeds reforestation, so this process is a net source of carbon in the atmosphere.

You can also see in Figure 5 that photosynthesis exceeds respiration by 0.4 Gt per year. More significantly, more CO_2 is going into the oceans than is coming out. Thus, the oceans are a net sink for carbon.

Finally, Figure 5 shows the impact of burning fossil fuel as 6.4 Gt per year. Given the large numbers we have been discussing, this may seem small. But the available sinks for carbon are removing only 3.8 Gt a year while deforestation is adding 1.6 Gt and fossil fuels add 6.4 Gt. So there is a net increase of 4.2 Gt per year in this scheme. This is leading to an increase in atmospheric CO_2 that is well documented and uncontested. The conflict in the climate change debate involves whether this rise in CO_2 is responsible for an observed increase in global temperatures and whether this is a problem.

Keeping the Temperature Constant: Homeostasis and Feedback/Forcing

We have covered the major physical processes of the atmosphere, but we have not yet considered how these processes and the biosphere interact with temperature. To begin this discussion, it is useful to consider how your body keeps its temperature constant at 37°C (98.6°F). Some of the mechanism for this is shown in Figure 6.

The process of keeping something constant in systems is referred to as *homeostasis*. In engineering language, the various processes involved are called negative *feedback* processes, and the language of climate often refers to these as *forcing*, negative and positive.

For body temperature there are several mechanisms that are used. When the body temperature gets too high, the first action is to dilate the blood vessels of the skin. This sends blood to the surface where heat can be radiated away. This also leads to sweating, and the evaporation of sweat from the skin cools the blood. These normally return body

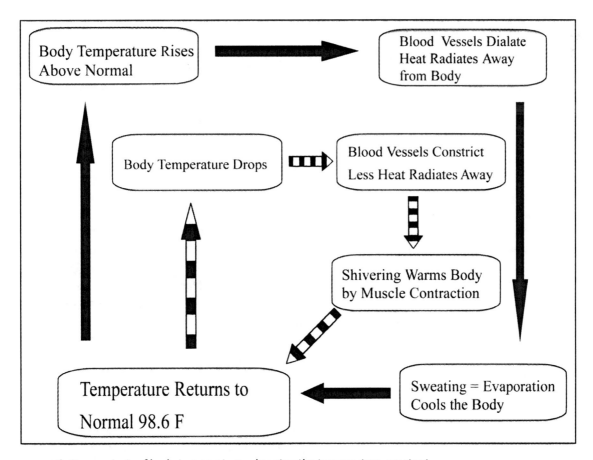

FIGURE 6 Homeostasis of body temperature—keeping the temperature constant.

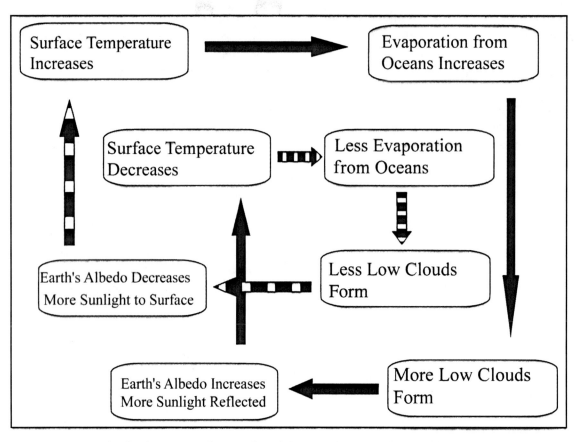

FIGURE 7 Negative feedback system in climate. The solid arrows denote cooling, and the dashed arrows denote warming.

temperature to normal. When the body temperature starts to drop, the process is reversed. The blood vessels in the skin contract, drawing the blood to the core. This reduces the radiation of heat and stops the sweating process. If this is not sufficient to maintain temperature, the body begins to shiver. This rapid contraction of muscles burns glucose, and the process of consuming calories releases heat. So shivering actually does warm up the core temperature. The sum of all these processes constitutes a complex negative feedback system that maintains core temperature for the body.

There are a variety of similar mechanisms that keep the global temperature comfortable for life on Earth. James Lovelock, author of *Gaia: A New Look at Life on Earth*, was among the first to recognize this. His work and the efforts of both his supporters and critics have identified a number of these feedback mechanisms. Some are physical and some are biological. Thus, Lovelock's hypothesis, that a combination of biological and physical mechanisms produces homeostasis, is now widely accepted. A few examples will be useful.

Clouds in the atmosphere increase the amount of incoming solar energy reflected back into space. So lots of clouds can have a cooling effect. (The actual effects of clouds are a bit more complicated and depend on the type of clouds—some clouds actually have a warming effect. But for this purpose, we will look only at the reflection of incoming energy.) Beginning in the top left of Figure 7, when the surface of the Earth gets warm, more water evaporates. This leads to more clouds, and they in turn cool the surface, returning it to the homeostatic condition. The inner loop of Figure 7 shows the opposite effect, when the surface cools below the homeostatic condition. Then less evaporation reduces clouds and results in surface warming.

Lovelock identified a biological mechanism that does something similar. Plankton in the oceans produce a gas called dimethyl sulfide as part of their response to sunshine. The more sunshine there is, the more of this gas is produced. The gas moves into the atmosphere where it reacts with air and sunlight to form sulfate aerosol particles. It has been shown that a large proportion of all clouds formed have these sulfate aerosol particles at the center of the water droplets. Without the aerosol particles, the water vapor does not condense as easily to form clouds. Thus, sunshine on the oceans helps produce clouds, which reduce sunshine on the ocean. This in turn reduces the production of dimethyl sulfide, which reduces clouds. That causes the dimethyl sulfide production to increase again. This is a classic negative feedback process. There are many other examples, but these suffice to show how the system works.

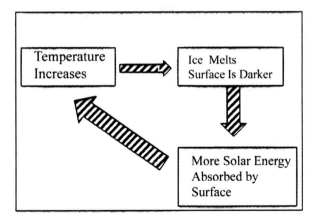

FIGURE 8 Ice positive feedback system.

There are also positive feedback processes. These are bad for climate because they tend to make systems unstable and cause wild swings in systems. One positive feedback system is shown in Figure 8. Here the warming of the surface melts ice, which increases the fraction of solar energy absorbed by the surface and decreases the fraction reflected to space. This warming melts more ice, and the process accelerates until there is no more ice. This process is at work today in the Arctic region where ice at the North Pole has decreased by 40 percent in recent decades and is expected to disappear completely in the next decade.

A biological example of positive feedback is also occurring in the far north. Alaska has warmed by 9°F in the past twenty years. This has caused the permafrost to melt and released methane and carbon dioxide trapped in the ice. These greenhouse gases increase warming and melt more ice. The thawed soil also contains a lot of biological material that becomes food for microorganisms and is converted to CO_2. The warmer temperatures speed up the decomposition of this material, which in turn makes the temperature even warmer. The positive feedback of temperature and biological activity causes the process to accelerate, as shown in Figure 9.

Sources of Greenhouse Gases

The two greenhouse gases that are most important and difficult to control are CO_2 and methane. It is informative to see where these come from, because any decision to control or reduce these gases requires a clear understanding of where to look for solutions.

We have already seen that deforestation is a source of CO_2 emissions, as shown in Figure 5, which shows the carbon cycle. Major action to decrease deforestation or a major reforestation program could shift the balance such that forests became a carbon sink instead of the source of 1 billion tons of extra atmospheric carbon each year.

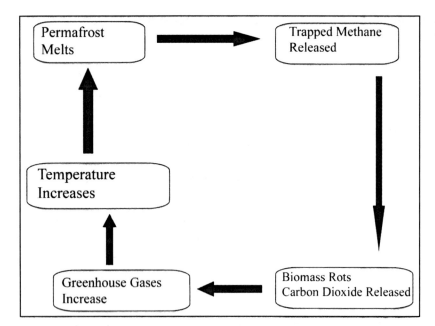

FIGURE 9 Permafrost positive feedback.

The other three major sources shown in Figure 10 are power stations, transportation, and industry.[8] Some industries such as the chemical industry have made major reductions in emissions. These changes were motivated by profit: the savings in fuel have more than made up for the cost of equipment and process changes. In order to understand the other two sources, we need to examine the technologies used in generating electricity and in transportation.

Generating electricity using coal, oil, and other fuels involves burning the fuel to make steam. The molecular structure and names of typical fuels are described in Appendix 2. The steam then turns a turbine to produce electricity.[9] This process is shown in the Figure 11. The amount of energy produced by each fuel relative to the amount of carbon dioxide released can be calculated using the methods in Appendix 3. The bottom line of these calculations is that coal is the worst possible fuel from the standpoint of climate. Oil is somewhat better. Natural gas releases only about half the carbon dioxide per unit heat as coal.

Scientists have determined that the process shown in Figure 11 is limited by the laws of nature to a maximum efficiency around 40 percent. The actual efficiency is always less than this and depends on the difference in temperature of the water entering the boiler and coming out as steam. Thus, it is necessary to cool the water after

---

8. Duke University, Google Earth, and Environmental Problems and Solutions, "Carbon Dioxide," https://sites.duke.edu/tlge_sss29/carbon-dioxide-emissions/carbon-dioxide/; PBS Frontline, "Heat: Sources of World's CO2 Emissions," October 21, 2008, https://web.archive.org/web/20081025083323/http://www.pbs.org:80/wgbh/pages/frontline/heat/etc/worldco2.html.

9. Tennessee Valley Authority, "How a Coal Plant Works," accessed 2012, www.tva.com/Energy/Our-Power-System/Coal/How-a-Coal-Plant-Works.

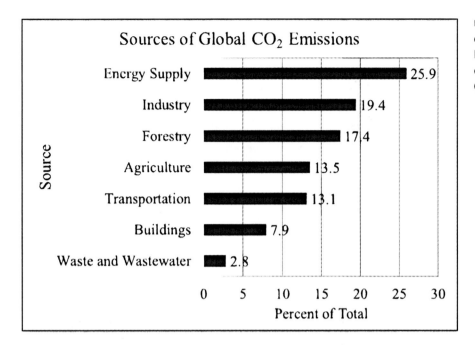

FIGURE 10 Sources of carbon dioxide. (Adapted from PBS Frontline, "Heat: Sources of World's CO2 Emissions," October 21, 2008.)

the turbine, and this produces lots of "waste heat." The waste heat usually ends up in rivers or lakes. Actual power plants rarely get more than 35 percent of the energy of the fuel burned converted to electricity. Nuclear power plants use the same process, replacing the boiler with a nuclear reactor.

An alternate method of generating electricity shown in Figure 12 is much more efficient and can reach efficiencies over 90 percent if well designed and sited.[10] This system replaces the boiler with a turbine engine similar to those in jet aircraft. The natural gas fuel is injected into the gas turbine engine to spin the turbine and make electricity by turning Generator 1. The exhaust gas from the turbine is still very hot, possibly 1,000°F. This heat is used to boil water in a traditional boiler and to spin a second turbine. This generates electricity in Generator 2. The steam leaving the steam turbine is still quite hot, and this heat can be used as process steam in a variety of industrial processes or to make hot water.

The process shown here is both theoretically and practically more efficient than a traditional boiler because it uses two different generators to get the most from the fuel burned. These combined cycle systems approach 60 percent overall efficiency in turning fuel into electricity. A further efficiency is obtained if the waste heat in the process steam can be used for something that would otherwise require burning fuel, such as hot water for showers in a dorm. This is shown as Hot Water User in the figure. The term for this is combined heat and power, or *cogeneration*. Cogeneration plants are built where there is a need for hot water for industrial or other processes. Several hundred U.S. colleges and universities have these plants to provide not only electricity

---

10. Kable Intelligence, "San Joaquin Valley Energy Center, CA, United States of America," Power-Technology.com, last updated 2017, https://web.archive.org/web/20170617213848/http://www.power-technology.com:80/projects/san_joaquin.

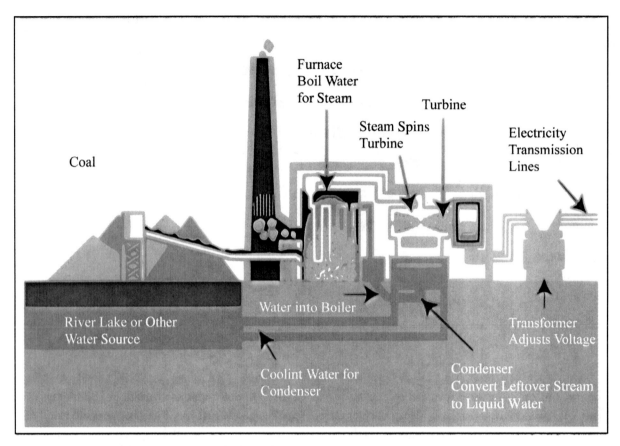

FIGURE 11 Traditional electrical generation process. (Adapted from Tennessee Valley Authority, n.d.; www.tva.com/power/coalart.htm.)

but also the hot water to wash students' clothes and dishes and to provide showers. Such systems often exceed 90 percent overall efficiency. Thus, replacing a traditional boiler with cogeneration offers as much as a 60 percent reduction of CO_2 emissions.

The other major source of fossil fuel emissions is in transportation. The use of gasoline to power cars and trucks is very inefficient indeed.

A typical gasoline internal combustion engine in a car uses less than 20 percent of the energy in the initial barrel of crude oil to drive the car. The remaining 80 percent is lost to heat and friction. Doubling the efficiency of cars would reduce the transportation emissions to half of their current values. This does not require any new technology. The relative efficiencies of various automotive technologies are shown in Figure 13.

Hybrid gasoline cars like the Toyota Prius are now readily available and cost effective. These systems effectively double the efficiency of cars by improving the tank to wheels efficiency. Electric vehicles are also now in commercial production, but their value depends on the source of the electricity. If it comes from coal, then they offer no real advantage for the climate problem. The data in Figure 13 assume that a gas-fired power plant is the source. However, if the source is renewable, such as wind or solar, electric vehicles reduce CO_2 emissions to zero.

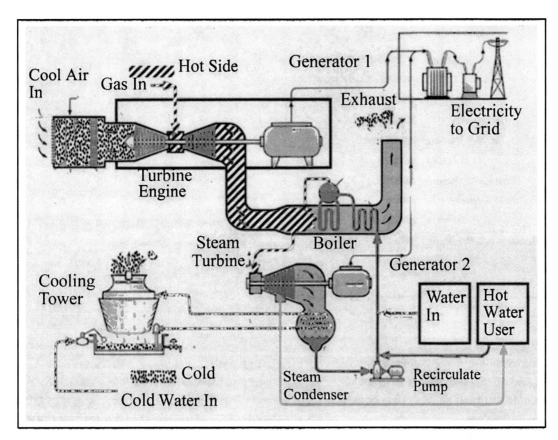

FIGURE 12 Cogeneration: combined heat and power generator. (Adapted from Kable Intelligence, "San Joaquin," www.power-technology.com/projects/san_joaquin/images/Combined3.jpg.)

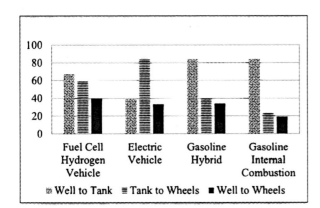

FIGURE 13 Well to wheels efficiency of cars. (Data from Toyota Motors.)

Fuel cell hydrogen vehicles (FCHV) are not yet commercially available. They have the potential to offer even higher efficiency. They can be fueled with hydrogen made using renewable electricity or potentially directly from solar energy. They can also be fueled by breaking natural gas down into hydrogen, as in the example in Figure 13. Both FCHV and electrical vehicles in Figure 13 are using fossil fuel as their ultimate power source. Their efficiency in terms of greenhouse gases rises to 100 percent if the electricity or hydrogen used come from solar, wind, or other non-fossil sources.

Another way to save energy in the transportation sector is to make more use of trains rather than trucks. Trains are often advertised on television as

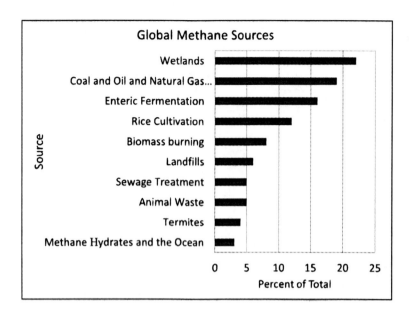

FIGURE 14 Sources of methane. (Adapted from Augenbraun et al. [NASA], 1997.)

being able to move a pound of products 400 miles on a gallon of fuel. Increased use of rail to move freight and light rail to replace cars offers considerable savings in emissions.

Methane is the other major greenhouse gas, and controlling methane presents a challenge. Methane, which is natural gas, has a number of sources, as shown in Figure 14.[11] A considerable amount is lost to leakage in the production and transportation of natural gas and other petroleum products. But the largest overall sources are related to agriculture. Rice cultivation involves the large production of methane by natural bacteria in rice paddies. The meat and dairy industries are also major sources (the enteric fermentation in Figure 14 refers to belching and farting cows). These sources are much harder to control than the major sources of CO_2. Research is underway to find ways to grow rice and raise cattle with lower methane emissions. Burning biomass, of which deforestation is a major component, also contributes to this gas. Wetlands also have a major impact on methane emissions, and it is not clear that anything can change this.

Black carbon particles are another air pollutant that warms the earth. This is a fancy name for soot, which is produced by burning fuel. Diesel engines are one major source, but inefficient use of cooking fuel in the third world is another major contributor to this pollution. Controlling black carbon emissions may offer one of the least expensive and most practical short-term actions to reduce climate change.

Conclusion

Several factors should be clear from the previous discussion.

1. We have a good understanding of the sources of greenhouse gases.
2. We have the technology to make large reductions in emissions of CO_2.

11. H. Augenbraun, E. Matthews, and D. Sarma, "The Global Methane Cycle," Education: Global Methane Inventory, GISS Institute on Climate and Planets, NASA Goddard Space Flight Center, 1997 (last updated August 2, 2010), https://web.archive.org/web/20161223225005/http://icp.giss.nasa.gov/education/methane/intro/cycle.html.

3. These reductions will generally reduce the operating costs of everything that uses energy, from driving cars to lighting and heating homes.
4. The use of these efficient technologies produces displacements in the economy, particularly in the coal industry.
5. The capital costs of replacing infrastructure and constructing alternative energy systems are significant barriers to implementation.

Thus, the ultimate answers to any problem with greenhouse emissions are political and economic rather than scientific, though new scientific and engineering advances in energy storage, biofuels, and solar energy conversion can change the economic picture in important ways.

What Can Be Done about Greenhouse Gases?

There are a number of strategies that have been proposed to deal with the threat of climate change due to greenhouse emissions. Some of these are inexpensive and even profitable. Some are expensive and may not even work. The list here is not exhaustive but rather is intended to start you thinking about solutions.

1. Reduce emissions through a tax on carbon.
2. Reduce emissions through international and national cap and trade programs.
3. Reduce emissions using caps and penalties for excess emissions.
4. Reduce emissions through increased energy efficiency to reduce total fuel use. This includes efficiency improvements in transportation, housing, industry, and energy production. This would be a natural outcome of any of the first three actions, but there are alternate ways to achieve these goals, including mandatory efficiency standards independent of caps or taxes.
5. Reduce methane emissions due to loss in refining and transportation.
6. Reduce agricultural emissions, including the methane generated by cows used for meat and dairy, from rice paddies, and from use of fertilizer.
7. Reduce emissions by changing land use and reducing deforestation.
8. Increase the removal of gases through planting trees and preventing deforestation.
9. Increase the removal of gases through changes in land use and agriculture.
10. Sequester CO_2 from major point source emitters (such as power plants).
11. Switch from fossil fuel to biomass-derived fuel.
12. Build more wind, hydroelectric, and solar electrical power generation capacity.
13. Build more nuclear power plants.
14. Reduce solar energy input through geoengineering in space or the stratosphere.
15. Reduce black carbon particulate emissions from transportation and cooking fires.

Goals for Minimizing Climate Change

There is no scientific consensus on how much temperature can rise before irreparable harm is caused to the climate and to human societies. The amount of warming that can be tolerated by societies depends in large part on how sensitive their location is to sea level rise, to increases in desert area, or to flooding, loss of agricultural land, and so on. Some nations may disappear with only 1.5°C warming, but others may prosper.

Global climate models estimate the changes that may occur with varying levels of greenhouse gases in the future. These models have been tested using historical data for the past century, and they reproduce the observed warming reasonably well. The danger inherent in these models is that they assume all climate processes will change in a linear and predictable way. They do not include, and cannot include, the unexpected changes that might occur. Climate data show that global temperatures changed quite rapidly during certain prehistoric events. The source of these sudden, drastic changes is unknown, but their existence causes more conservative scientists to suggest that model results may not include all factors needed to predict climate if greenhouse gases rise above the present level.

POLITICAL BACKGROUND: INTERNATIONAL ENVIRONMENTAL TREATIES

The process of developing international treaties to address any issue is a difficult one. The Montreal Protocol was negotiated in the late 1980s to control chlorofluorocarbon emissions that were destroying the stratospheric ozone layer. The Long Range Transport Air Pollution Treaty was a regional treaty for acid precipitation that was negotiated to cover Europe. Both of these treaties required multiyear efforts. In both cases, the negotiations began while there was some question about the scientific basis and risks associated with the pollution. The period of negotiations included research to clarify the problem, and this research ultimately convinced the majority of nations that a treaty was necessary. Only then were the negotiations successful.

The actual negotiations, such as those that led to the Montreal Protocol and the ongoing climate negotiations that constitute this game, occur at several levels. There is usually a technical level of work by committees of experts, scientists, economists, and others, who draft potential treaties for the political leaders to then debate. The treaty draft in Appendix 4 is an example of this type of work. The draft offers options and blank spaces that require political decisions. For example, the draft does not specify how many degrees of warming will be set as the upper limit or the actual amount of money that might be committed to help support the process (see, for example, section I.2 in Appendix 4).

Negotiations within the Conference of Parties process reach the top political level at conferences such as Copenhagen. The goal is for national leaders to make the decisions and reach compromises on contentious issues so that a treaty can be accepted by the majority. However, an agreement at Copenhagen would only be the first step in bringing the treaty into force. For democratic nations, treaties must be approved by each nation in their parliament or congress. In the case of the United States, a treaty requires approval by two-thirds majority of the Senate. So the fact that the president agrees to a treaty does not make it happen. It is just the first step.

Treaties normally include language that specifies they take effect only when the number of ratifying countries reaches a specific level, usually one-half to three-fourths of the countries. The details of this are another issue subject to negotiations. In the case of the climate treaty, the number of nations approving the treaty is less important than the amount of their emissions. A successful treaty will need to be ratified by the countries that contribute more than a majority of total global emissions. The three largest polluters are the United States, the People's Republic of China, and India. Thus, without ratification by all three of these, it is unlikely that a treaty would have any impact, no matter how many small countries sign it.

For the Kyoto Protocol on Climate Change it took a number of years to get the required number of nations to ratify this treaty before it could be brought into force. The United States representative to Kyoto agreed to the treaty, but it was never submitted to the Senate for ratification due to domestic political opposition. Ratification by a large majority increases political pressure on the remaining countries to also ratify the treaty. In some cases, further negotiations are required to make modifications to get a few holdouts to ratify.

Pollution Reduction Mechanisms

A variety of mechanisms exist for reducing pollution. This game does not allow time to enter into the specific mechanisms that will be used during the negotiations. In most cases these mechanisms have been applied primarily at the national level to meet their emission goals. Due to the national differences in economic models, different approaches work better in different countries. A country that has public ownership of their electric utilities will necessarily use a different approach than one in which the utilities are privately owned. Some mechanism can also be applied internationally.

Command and Control

The command and control approach is one in which the government mandates specific actions and requires all industries and individuals to comply or face civil or criminal penalties. The mandate can be for the use of a specific technology. For example, a government could require that all coal-fired boilers over a certain capacity be fitted with carbon sequestration technology by a specific date.

A second approach to command and control requires the use of the *best available technology*. This creates a moving target of regulations in which a specific class of polluters is required to adopt whatever is best, and as the technology improves the mandate changes.

A third approach is to mandate specific emissions levels rather than the technology to get there. This approach allows each individual plant to determine the most cost-effective way to reach the mandated limit. The specific form of the limit or cap on emissions can be in the form of annual totals, daily totals, maximum per unit of time, or peak maximum. Thus, a plant could be allowed to emit 1,000 tons of CO_2 per week, 300,000 tons per year, or no more than 100 kilograms per hour, or to have emissions that never exceed 100 kilograms per hour. This approach to regulation requires more monitoring and reporting than the requirement for specific technology, but it is also more flexible.

Polluter Pays

A second approach to pollution control is simply to establish a cost per unit of pollution that must be paid by the polluter. This creates an economic incentive to reduce pollution and can be a source of government revenue to cover the costs of pollution. A power plant might be assessed a charge of 10 Euro per kilogram of CO_2. This approach also requires monitoring to ensure compliance. It has the advantage of allowing economic forces to act on pollution, but it also makes it possible for companies with large profits to simply pass the charges along to their customers and continue polluting. The economic forces to remove pollution are only felt strongly if there is a competing source for the same product or service that pollutes less and can thus sell at a lower cost.

An alternative approach is a carbon tax. This would be levied on the production of fossil fuels based on the total carbon emissions when they are burned. A tax per ton of coal and per barrel of oil would increase the cost of fossil fuel energy, and the cost would be passed through to the end user of the products. In this way, the more energy used to make a product, the greater the cost. Thus, a product that is less energy intensive will be less expensive. This is a much easier system to implement than a tax on emissions for CO_2.

Cap and Trade

The cap and trade approach, popularized by the Reagan administration in the United States, combines some of the features of the two preceding approaches. The cap part is a mandated limit on total emissions of a specific pollutant for a nation or even for the entire globe. The cap can be gradually reduced over time to provide additional reductions. Once the total emission cap is determined, the right to emit the pollutant is allocated to all polluters on some basis. This takes the form of emission credits that allow a specific amount of pollution.

This allocation process must be perceived as reasonably fair to be politically viable. Usually this is done based on some fraction of current emissions. A market is then established in which emission credits can be bought and sold. For example, a company is allowed to emit 10,000 tons of CO_2 but has improved their efficiency so that they can produce the same products with only 8,000 tons of emissions; they can sell their credits for 2,000 tons on the open market. A polluter who finds it less expensive to buy those 2,000 tons of emission rights than to install equipment to reduce 2,000 tons of CO_2 emission can purchase the credits as a way to stay within his limits.

The value of the emission rights will fluctuate depending on supply and demand. If the caps are gradually reduced, the cost of emission rights will rise as they become scarce. This will make the installation of pollution control equipment more cost effective than buying emission rights for more and more polluters.

Compliance

Treaties between sovereign states are by their very nature voluntary. Compliance cannot be compelled easily, and public and political pressure are often the only tools to ensure compliance. Trade sanctions and tariffs could also be used to ensure compliance, but these have not generally figured in these negotiations. For any nation to comply with an international agreement, there must be general agreement that the principles behind the treaty are sound, that the mechanisms of the treaty are fair, and that there are reasonable mechanisms for resolving the inevitable conflicts that will arise.

Another aspect of international treaties is the penalty for failure to meet the obligations of the treaty once a nation has signed it. The exact details of this are beyond the scope of what can be addressed in the game. Various mechanisms are available. The World Trade Organization (WTO) is a body that resolves disputes related to various trade treaties. The International Court of Justice in The Hague, also known as the World Court, provides a venue for nations to seek damages from other nations. If a legal treaty were ratified that required specific reductions in carbon dioxide emissions, the countries that suffered damage could seek money from the countries that failed to meet their obligations.

It may also be possible for countries damaged by climate change to seek damages under existing UN treaties that make nations responsible for pollution they generate. However, this has not yet been tested for greenhouse gases. If a group of countries has agreed to reduce their emissions, they could also impose import tariffs on products produced in countries that have failed to join the group and reduce their emissions, so long as these tariffs do not violate their trade treaties within the WTO.

GLOSSARY AND GUIDE TO ABBREVIATIONS

CDM	Clean development mechanism
CFCs	Chlorofluorocarbons
CO2	Carbon dioxide
COP	Conference of Parties
EU	European Union
Gt	Gigatons
HCFC	Hydrochlorofluorocarbons
IPCC	Intergovernmental Panel on Climate Change
IR	Infrared
JI	Joint implementation
K	Kelvin
LRTAP	Long-range transboundary air pollution
mol	Mole(s)
NGO	Nongovernmental organization
O3	Ozone
U.K.	United Kingdom
UN	United Nations
UNFCC	UN Framework Convention on Climate Change
OPEC	Organization of the Petroleum Exporting Countries
ppm	parts per million
UV	ultraviolet
W	Watts
WFA	Western Fuels Association
WTO	World Trade Organization

2

The Game

MAJOR ISSUES FOR DEBATE

Each player has victory objectives built tearound the issues described here. In most cases players have strong feelings about one or two objectives and are indeterminate on the rest. Your arguments in the class should be developed to convince the players who may be indeterminate on your key issues to vote for your positions.

The debate at Copenhagen covered a massive number of issues and went on for several weeks. This game will only address a few of the most important issues. Appendix 4 is an annotated draft of a proposal by the Danish hosts. Note that there are blanks that must be filled in based on what the assembled nations can agree on. Even with the blanks, this document angered many of the delegates for various reasons. Section numbers are provided to guide you to the relevant portions of the Danish draft. The most significant parts of the draft are highlighted in **bold** type so you can find them easily. The remaining text of the draft, while significant, is not something that most of you will consider in detail at this meeting. In developing your treaty proposal, it is not necessary to write a major document. All you need to do is specify the answer to five points. Table 2 has been provided below to clarify this.

Issue 1: Limits on Greenhouse Gases

The nations at Copenhagen vary in their opinions, from wanting a very restrictive treaty aimed at a 1.5 degree limit corresponding to negative net emissions by 2050 to no limit at all. In addition to a long-term goal for 2050, it is also important to agree on a short-term goal. For purposes of this game, the short-term goal has been set as 2020, but the original Danish draft left this year open for debate. See the draft treaty in Appendix 4 (sections I-3, III-7, V-20, VI-30, and Attachments). Many scientists have suggested 400 ppm CO_2 equivalents and 2°C warming would be a good target. Others propose that 1.5°C warming, while more demanding, is required to avoid significant problems. Some propose that global CO_2 should

TABLE 2 Treaty characteristics for debate

Issue	Proposed Treaty, Short Term 2020	Proposed Treaty, Long Term 2050
Limit on temperature increase or reduction in greenhouse gases 2050 (see Figure 15 & Figure 16)		
Political or legal treaty?		
Transparency strong/weak		
Money for developing countries—energy development, adaptation, and remediation	$Billion/year	
Money for forest preservation	$Billion/year	

actually be reduced from present levels to less than 350 ppm. Most experts agree that by 2050 overall cuts of at least 80 percent are required to stabilize climate.

These are things to consider when drafting a treaty. A treaty that calls for holding temperature increases to 2°C can be assumed to require 20 percent short-term cuts and 90 percent long-term cuts for at least the major industrialized nations. A less restrictive treaty would require fewer reductions, and a more restrictive treaty will be more demanding.

For the purposes of this game, Figures 15 and 16 can be taken as the Intergovernmental Panel on Climate Change (IPCC) consensus. These are derived from the IPCC Climate Summary 2007.[1] Because the IPCC provides a range of values rather than single values, their numbers have been simplified to make it easier to play the game. But remember there is a considerable uncertainty in these values.

When writing a treaty to control climate change, the language can either require a specific maximum warming or a specific target for overall greenhouse gas reductions by 2050. The use of a temperature limit would require that adjustments in emissions be made as data improve, so the actual reductions might be more or less than the estimates in 2009. If a treaty specifies a specific percentage for global greenhouse gas reductions,

---

1. The full report is available online: Intergovernmental Panel on Climate Change, *Climate Change 2007: Synthesis Report. Contribution of Working Groups I, II and III to the Fourth Assessment Report of the Intergovernmental Panel on Climate Change*, ed. R. K. Pachauri and A. Reisinger (Geneva: IPCC, 2007), www.ipcc.ch/publications_and_data/publications_ipcc_fourth_assessment_report_synthesis_report.htm.

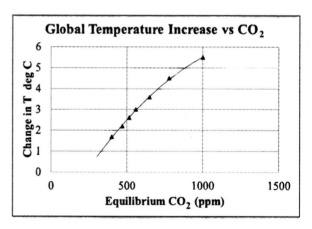

FIGURE 15 Estimated relationship between final equilibrium carbon dioxide levels and global temperature. (Adapted from IPCC, Climate Change 2007: Synthesis Report, table 5.1.)

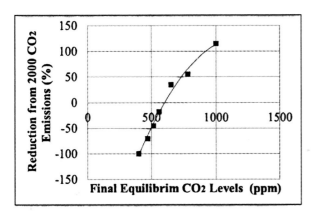

FIGURE 16 Percentage reduction in global greenhouse gas emissions required to reach various equilibrium concentrations. (Adapted from IPCC, Climate Change 2007: Synthesis Report, table 5.1.)

the actual temperature may change more or less than predicted. In either case, a treaty needs to specify how the cuts in emissions are allocated among countries and how they are phased in. Interim targets for dates such as 2020 need to be included to ensure that nations begin reductions immediately.

A treaty with specific targets for reduction is easier to adjust for varying national situations but will not necessarily produce the desired result. Although the People's Republic of China and India will soon be the largest producers of total greenhouse gases, their per capita emissions are tiny when compared with the United States. A successful treaty will need to allow continued development of some economies without destroying the economies of the developed nations.

The IPCC report shows ranges of values, which we have converted to average values to provide a simple way to discuss them in the game. The actual values have a large uncertainty (up to 50 percent), and that uncertainty may be an issue for some players' roles in the game.

Issue 2: Nature of the Agreement *[handwritten: Legal or polit.]*

The treaty must address the issue of a *legal or political treaty*. The strongest type of treaty would be legally binding on the signers and could be enforced by international courts. A political treaty, on the other hand, is a commitment without consequences for failure to comply. A nation that signs a political treaty and fails to comply will face international publicity and criticism but no other consequences.

A legal treaty must have specific consequences for failure to meet the obligations of the treaty. These consequences could apply to countries that do not sign the treaty as well. Countries that sign the treaty could impose a tariff of some amount on all imports from countries that either fail to sign the treaty or fail to meet their obligations to reduce emissions. For example, the treaty could specify a tariff of 5 percent on the value of all imports from these countries into members of the treaty organization. Negotiations for a legal treaty would include setting the amount of such a tariff. The tariff income could pay for remediation to counter the effects of the emissions from the noncompliant countries. The effectiveness of such an approach depends on having a large group of countries in the treaty so that their economic power is sufficient to make noncompliance expensive. (See preamble to the Copenhagen Agreement draft.)

Issue 3: Transparency *[handwritten: How much reporting will be required]*

Transparency covers all aspects of the agreement. The leaked Danish draft includes strong transparency and would require detailed reporting of emissions, inventory of forests and land use, and strong accounting of all money donated and received for mitigation and emission reduction. This accounting would be monitored by an independent body.

Weak transparency eliminates these safeguards, and that may be favored by nations who want to protect the internal workings of their countries from outside scrutiny. Weak transparency also makes it easier to cheat in a variety of ways. An intermediate level of transparency would require only aggregate values rather than detailed inventories. It is also possible to have no reporting at all,

THE GAME 155

but this would represent a very weak and unenforceable treaty. (See section V-24.)

Issue 4: Money Transfer from Developed to Developing Countries
[margin note: Funding]

Money transfers would establish a fund to mitigate the damage caused by climate change and to allow poor, developing countries to increase their energy availability per capita without increasing greenhouse emissions. Developing countries will probably not sign an agreement without significant funding in this category. (See section V-19-22.)

Issue 5: Money to Protect Forests
Forest protection may be treated as a subcategory under mitigation, but the recipients may not be the same countries as in issue 4. The Amazon rainforest is one of the largest potential sinks for carbon, and Indonesia and several other countries fall into this category as well. Deforestation not only releases CO_2 but also destroys a carbon sink. There is considerable interest in steering development in these countries away from forest destruction, but this requires financial support from the developed world. This may be included in the funding under issue 4, or it could be a separate fund. (See section III-12.)

Issue 6: Remediation
It is clear that no amount of action in 2009 will prevent some increase in CO_2 and global temperatures. These changes will raise sea levels, forcing relocation of significant populations. The distribution of and access to fresh water will change. Species extinctions will occur in many areas. Land for agriculture may become unusable due to temperature, or lack rain or fresh water for irrigation. Some countries are particularly vulnerable to these changes, including Egypt, Bangladesh, India, and China. It has been estimated that as many as 300 million people will need to be relocated by 2050. Money will be required to pay for remediation of these problems. The amount, distribution, and transparency of this process are an important issues. Players must include arguments for their position on these issues in their presentations.

RULES AND PROCEDURES

Conference Protocol
The Copenhagen meeting will be chaired by the Prime Minister of Denmark, the host country. The normal rules of parliamentary procedure can be employed to run the actual meeting. The chair should work to ensure good order and make certain that all countries have adequate time to present their positions. The participation of the nongovernmental organizations (NGOs) is at the pleasure of the chair. Every NGO must be allowed to make at least one presentation. After this, the chair will control who is allowed to speak. If the NGOs become overly disruptive, the chair may have the meeting closed for country representatives only.

Private Meetings
Negotiations for a treaty are often best conducted in private meetings between small groups of countries. Any country may request a recess of the formal meeting so that they can hold private faction meetings or private negotiating sessions, but only the chair may grant this request during regular sessions. Naturally, players may try to meet on their own outside the regular class meetings.

At any point in a session, the chair may convene a brief "Friends of the Chair" meeting for selected nations to consult in private. Countries excluded from these meetings are free to intrude on these sessions if they can.

Representatives of NGOs and journalists should attempt to observe these private meetings as well, but they but may be excluded by the chair.

Voting at the Conference
Only country representatives have a vote in the meeting.

It is not possible to call for a binding vote on each issue one at a time because the vote on one issue will be contingent on the vote for another. The chair may call for a "straw vote" on individual issues to get a sense the conditions that must be met to pass a specific aspect of the treaty.

Rather than voting on individual issues, groups of players must caucus together to draft treaties that address all the major issues. These treaty drafts may be amended.

Once the chair is confident that a particular treaty draft stands a good chance of passage, the chair should call for a vote. Approval requires a majority of country representatives voting in favor of the treaty. If a majority votes "yes," the treaty is approved.

However, everyone must recognize that only the nations that vote for the treaty are bound to comply with its conditions. If the United States, China, or India fail to approve the treaty, it will have little global impact. Thus, even if a treaty is passed, if these countries do not support it, the conference will be considered a failure.

After the Conference: National Approval

Country representatives are constrained by their governments. This means that no matter how much a player representing a country may wish to vote for a particular treaty draft, they cannot if it contradicts their victory objectives. If players ignore these objectives, the Gamemaster will cancel their votes, explaining that the treaty failed to be ratified by their home governments (the U.S. Senate in the United States, the Politburo in China, the king in Saudi Arabia, and national parliaments elsewhere).

OUTLINE OF THE GAME

Game Session 1: Opening Session

1. The Danish prime minister opens the session with a call for action to reduce climate change.
2. The IPCC outlines the evidence that human activity is changing the climate in potentially dangerous ways.
3. Climate skeptics will then respond with their own presentation.
4. After this, individual countries can introduce themselves and begin to present their specific concerns and challenge any aspect of the science they wish by presenting contrary evidence.

Game Sessions 2 (and 3): General Debate and Approval of a Treaty

1. Denmark briefly reviews its proposed draft treaty.
2. All players have the opportunity to respond—perhaps with draft treaties of their own.
3. Small private meetings and straw polls are likely to be useful.
4. The chair calls for a vote on a draft treaty. IPCC and NGO delegates do not vote.

Remember that only nations that approve the treaty are bound by it. As was the case in Kyoto, a treaty may go into effect if a 55 percent of nations agree to it and they represent at least 55 percent of total global emissions. This means that no treaty can take effect without the approval of the United States, China, and India.

Before the Final Session

In advance of the final game session, all indeterminate players must send the Gamemaster their positions on the key issues outlined on Table 2.
1. Denmark should formally propose positions on all the issues as a draft treaty.
2. Each faction should develop positions so they know what they want out of the treaty.

Debriefing

What actually happened, and how well did it work? What are the prospects for the future?

ASSIGNMENTS

Each student will normally write one paper for this game. Not every student in a faction may have identical objectives, and some may have secrets.

Each student should write arguments that provide evidence to support their *own* objectives. Within factions, it is important to divide the research and writing so that each student focuses on a limited number of issues. Thus, one student from the European Union might write only on the philosophical aspects of the argument and another on the scientific issues. A third might address the economic issues. This will allow for more depth in the writing and avoid superficial papers that all make the same arguments. The most important aspect of all papers is that they provide evidence to back up their arguments. It is important that you provide specific references for the factual information you use in your paper. The evidence you use must be something known in 2009.

The nature of the paper will depend on your role and the specific instructions given by your instructor. Because much of the data are graphical or numerical, you may want to prepare a PowerPoint presentation as well. In some cases, instructors may have you provide an outline of your major points and the references that support your points rather than write a formal essay.

3
Roles and Factions

INDETERMINATE COUNTRIES

The indeterminate countries must be convinced to sign a treaty (or dissuaded from it) by those on the opposing sides of the issue. Many of them would like a treaty but have conditions that must be met to sign.

United States, Australia. These two developed countries have some of the highest per-capita and per $GDP output of greenhouse gases. Both countries have thus far refused to agree to an international treaty.

Brazil, Russia, India, People's Republic of China (BRIC). These are large countries with rapid development. China will overtake the United States as the largest emitter of carbon dioxide as soon as 2010. China and India have over a billion people each, many of whom seek to attain middle-class status with cars, refrigerators, and other energy-intensive devices. The rapid growth of these countries makes their emissions the greatest challenge for the future. Brazil is home to the Amazon Rainforest. Cutting the trees of the Amazon to meet the lumber needs of the developed world and burning its forests to raise soybeans and cattle for export to developed nations has four global impacts. The burning forests release carbon dioxide, the loss of the forests removes a major sink for carbon, the forests change the albedo of the earth, potentially leading to more warming, and the forests are home to many important species.

Russia is grouped with the other BRIC countries but has some unique issues. The massive contraction of the Russian economy has produced a reduction of emissions relative to the Soviet era, unlike the cases of China and India. Furthermore, as a major producer of oil and natural gas as well as coal, Russia is an important contributor to the problem, and is heavily dependent on oil and gas exports to support its economy. The Russians are, therefore, sympathetic to the reluctance of energy producers to see their income reduced by reducing the use of fossil fuels.

Journalists. Members of this group, who may support a treaty may also be present and try to build support for a particular type of treaty. They can do this by what they decide to report and the questions they raise about a treaty. The media are critically important in building support for or opposition to a treaty. The country representatives are quite sensitive to this, and no democratic country can take actions without public support. Journalists are the primary people who shape what the people see and hear about the issue.

GREENS: COUNTRIES THAT WANT THE STRONGEST POSSIBLE TREATY

European Union, Norway. The EU nations and Norway have a strong history of working together to solve environmental problems, beginning with the 1979 Long-range Transboundary Air Pollution (LRTAP) treaty. This treaty has been strengthened to include greenhouse emissions as part of the EU's acceptance of the Kyoto Treaty. The EU is leading the drive for international regulation of all greenhouse gases. Norway is a major oil-exporting country. However, Norway has one of the strongest records for environmental protection of any nation and strongly supports a treaty to reduce the impact of climate change.

Maldives, Bangladesh, Tanzania, Somalia, Marshall Islands. These less developed countries have very little carbon impact on the world, but they are at highest risk for the consequences of global warming. The Marshall Islands are, on average, only three feet above mean sea level, so the predicted three-foot rise in the next century will essentially eradicate the country from the map. Bangladesh suffers from terrible flooding over a large fraction of its land area, so rising seawater will increase the crowding in this already crowded country, reducing usable land by as much as 50 percent and reducing access to fresh water which is already in short supply. Some estimates suggest that Bangladesh will need to relocate up to 30 million people in the next few decades. Tanzania relies on melting glaciers in the mountains for most of its water, and these glaciers are expected to disappear within twenty to forty years—which could mean the end of agriculture in much of the country.

Intergovernmental Panel on Climate Change. This UN-sponsored consortium of scientists meets regularly to examine the evidence for global warming and to issue periodic reports. These reports are based on an overall consensus of the scientists from around the world. Because scientists often disagree, there is considerable controversy when putting these reports together, and political considerations sometimes interfere with the science. Also, scientists rarely make unequivocal statements—they tend to talk in terms of probabilities rather than certainties. This provides ammunition for their critics. The IPCC representatives will try to present the scientific evidence in as clear and convincing a way possible. They have won the Nobel Prize for their work, so they have to be taken seriously. Still, the climate change deniers will challenge them at every turn. The IPCC will gain support from the International Friends of the Earth and the Pew Foundation for Global Climate Change in presenting their data. The Green movement has criticized the IPCC for being too conservative owing to its need for consensus, a charge that is supported by the fact that recent warming is greater than the highest IPCC predictions.

Nongovernmental organizations (NGOs). The International Friends of the Earth, the Pew Foundation for Global Climate Change, the European Green Party, and Greenpeace will press for strong and immediate action. NGOs have historically applied considerable pressure to the countries and have played an important role in all the climate meetings. They reinforce the scientific data and build pressure on the countries. They also may engage in demonstrations, make posters, and influence the media. The Copenhagen meeting has attracted about 35,000 NGOs, journalists, and observers.

Journalists. Representatives of climate-friendly news outlets may be present and will try to build public sentiment in favor of a treaty. Public pressure is the key to convincing the nations of the world to take action, and journalists have the power to focus public attention on the issues. They shape public opinion by what they chose to report and the way they shape their reports (spin).

OPPONENTS OF ANY TREATY

Saudi Arabia, Venezuela, Iran. These oil-producing countries are seriously concerned that any actions to reduce the burning of fossil fuel will diminish their income from petroleum exports. Their economies and the political stability of their ruling factions depend on the constant flow of money to provide public services. Any actions to reduce emissions will increase the price of energy and lead to reduced use. They have a deep fear that this will destabilize their countries, especially those in the Persian Gulf. This group is joined by NGOs— the Center for the Study of CO_2 and Climate and the Western Fuels Association (WFA)—that have spent a great deal of money supporting the continued use of coal and oil. Coal is the worst possible fuel from the prospective of global warming. The WFA has supported paid research to show the potential benefits of a high CO_2 world and to minimize the dangers. They also have considerable influence in the U.S. Congress due to the importance of coal mining in a number of states. Russia is not part of the Organization of the Petroleum Exporting Countries (OPEC), but the Russian economy is heavily dependent on exports of oil and natural gas. So the Russians may be sympathetic to the OPEC position during the conference.

Nongovernmental organizations (NGOs). The Western Fuels Association and the Center for the Study of CO_2 and Climate are two industry-funded groups with considerable clout in the U.S. political process. They will no doubt attempt to weaken U.S. participation, without which no agreement will have a significant impact on the problem. They will bring the "climate denier" point of view to the arguments. These NGOs and others have played a major role in raising questions about climate change and nurturing opposition in the U.S. Senate. They can also influence the media by spinning the science to shape the way the media report the issue.

Journalists. Publications that oppose a treaty may also be present and will try to stir up public opinion against a treaty. The ability of journalists and media outlets to cast doubt on climate science has been a significant factor in the failure of past meetings and has played an important role in the reluctance of national governments to take action on climate.

4
Core Texts and Supplemental Readings

The information in the appendices provides primary scientific background material. There are no specific books, but you will find many sources available on the Internet to provide material for developing your arguments.

It is important that you not use anything in the game that was not known in 2009. This does not mean that the publication dates must necessarily be before 2010 (because sometimes it takes time to get things published). It means no data or reports that reference data or information from after December 2009 can be used.

SUPPLEMENTAL SOURCES

It is important that you research the position of your country/faction to prepare empirical (evidence-based) arguments for your position. There are numerous websites with data on greenhouse gas emissions, gross domestic product, population, per capita income, and other aspects. All this information can be useful. Because the actual conference occurred relatively recently, you should be able to find information on the actual positions taken by your country. At the same time, you need to be careful not to use information that was not available in December 2009.

NationMaster is a website that allows you to select and graph a variety of parameters for all the nations of the world. For example, it is particularly useful for environmental comparisons such as CO_2 emissions: www.nationmaster.com/country-info/stats/Environment.

The Pew Center for Climate change has a number of reports that can be useful. The Pew Center has a reputation of being relatively unbiased, if there is such a thing, on this issue. Their reports are taken very seriously: www.pewclimate.org.

The International Panel on Climate Change (www.ipcc.ch) has a website where you can read the latest reports by this Nobel Prize winning group. The 2007 report at www.ipcc.ch/publications_and_data/ar4/syr/en/contents.html is quite useful.

The Center for the Study of CO_2 and Global Change (www.co2science.org) is one of the best sites for information that counters the IPCC position. Their website provides information on the upside of climate change and the benefits of added CO_2. This group receives major funding from energy industries.

GlobalWarming is a less aggressive opponent of action to slow climate change. Their site is located at www.globalwarming.org.

These are examples of just a few of the sites that may be useful for various factions. The key thing to remember is that every website has a point of view, and you need to be able to discern the site owner's agenda even if you do not recognize any of the people involved.

Acknowledgments

The work of Prof. Paula Lazrus and Prof. Nick Proctor is most gratefully acknowledged. Thanks are also extended to Prof. Gitte Schulz for reading the early manuscript of this game and providing many helpful suggestions. Also, the patience and hard work of the students of Chemistry 141 at Trinity College and in Prof. Schultz's first year seminar who played the first and second versions of this game are deeply appreciated. The example student papers for the game are edited versions of the work of these students.

Appendix 1. Greenhouse Gases

Much of what we know about atoms and molecules was discovered from the interaction of matter with light. The light we see with our eyes is only a small portion of the electromagnetic spectrum shown in Figure 17.

Electromagnetic radiation can be described in terms of the frequency of the waves, the wavelength of the waves, or the energy they carry. Frequency is the number of waves per second passing a fixed point and is reported in Hertz (Hz). If you were standing on the beach and could count the waves that came ashore in a fixed time, that would give you the frequency of the waves. Wavelength is the distance between wave crests. Looking out at the ocean, the wavelength would be the distance between waves.

Scientists use metric units for wavelength in meters and fractions or multiples of meters. Light moves much faster than ocean waves, and it is not possible for you to observe frequency or wavelength directly with your eyes. But the principle is the same. Wavelength and frequency are inversely related, so high frequency light has short wavelengths, and low frequency light has long wavelengths. Shorter wavelength, higher frequency waves have more energy that longer wavelength, lower frequency waves.

The energy of electromagnetic waves determines the kind of work the waves can do. We will see how this plays out in the interactions with matter.

The shortest wavelength electromagnetic radiation has wavelengths even smaller than the diameter of an atom. These are cosmic rays and gamma rays that result from radioactivity. At the other end of the spectrum are waves that are hundreds of miles long. These very long wavelength waves are used by the military to communicate with submarines.

The scale in Figure 17 is logarithmic. Every mark on the scale corresponds to waves that are ten times longer or shorter than the one next to it. The total range of the scale is on the order of the range from one penny to the national debt of the United States.

We will focus our attention on the range from infrared (IR) through ultraviolet (UV). This is the range that comes from the sun and is easiest to use for the study of atoms and molecules. The behavior of visible and IR light will be especially important to our discussion of climate change.

Atoms and molecules interact with light in a variety of ways. They can absorb light, in which case the energy of the light is transferred to the atom or molecule and does work on it. When UV light is absorbed, it moves an electron from one shell in the atom to a higher shell. The absorption of light allowed scientists to map to the energy levels of electrons in atoms and molecules. Ultimately, the observation of these interactions between light and atoms led to the field of quantum mechanics, which describes the behavior of electrons in all atoms and molecules.

Light in the IR part of the spectrum has a different effect when it is absorbed by molecules. Chemical bonds behave like springs, and the atoms in molecules vibrate constantly like weights connected by springs. When light in the IR spectrum is absorbed by a molecule, it causes the vibration to change if either of two things are true.

Case 1. Molecules with two or more different types of atoms have a more positive atom on one side and a more negative atom on the other. Bonds between atoms of different elements, whether C–H or C–O, have a small positive charge on one atom and a small negative charge on the other. These charges create a property called dipole moment. The dipole moment depends on the difference in

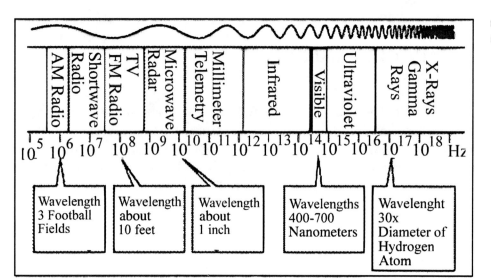

FIGURE 17 The electromagnetic spectrum.

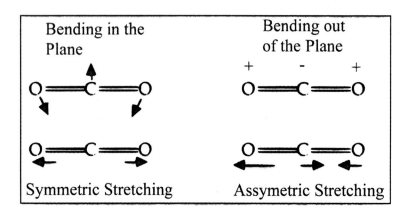

FIGURE 18 Vibrational modes for carbon dioxide.

charge and how far apart the charges are. These molecules tend to absorb IR radiation.

Case 2. Molecules with three or more atoms, even where all the atoms are the same as in ozone (O_3), will normally absorb some IR.

Molecules with only two identical atoms, such as like oxygen (O_2) and nitrogen (N_2), vibrate, but the vibrations do not change the dipole moment, and no IR is absorbed.

Molecules that absorb IR have specific frequencies of vibration, and the absorption spectrum of a molecule is like a fingerprint that allows the identification of molecules.

ABSORPTION OF IR RADIATION

We have discussed the fact that when IR light is absorbed the vibration of the molecule changes and only vibrations that change the dipole moment absorb IR light. Let us consider carbon dioxide (CO_2) in more detail, as its IR spectrum is important for the greenhouse effect. Figure 18 shows the ways in which the atoms in CO_2 can vibrate.

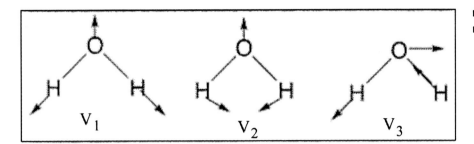

FIGURE 19 Vibrational modes for water.

If the two oxygen atoms move away from the carbon at the same time, the dipole moment of the molecule does not change (bottom left symmetric stretch). The two negative charges on opposite sides cancel each other out, no matter whether the bond is long or short. This vibration does not absorb IR radiation and is called IR inactive.

Another possible vibration has the carbon moving closer to the oxygen on one side, then moving closer to the oxygen on the other. This asymmetrical vibration moves the positive charge closer to one oxygen and makes the other end more negative. This vibration is IR active because the dipole changes due to the vibration. Two other vibrations are also IR active; in each of these, the carbon is displaced in such a way that the dipole moment changes.

Exercise

Consider the molecule H_2O. This is a bent molecule. What vibrations can this molecule have, and do they change the dipole moment? Which of the three modes in Figure 19 change the dipole moment?

Appendix 2. Chemicals in Fossil Fuels

One of the first tools of the early chemists was fire. We now understand that fire is the result of atoms and molecules reacting with oxygen from the air. The reactions of fuel with oxygen have provided power for humans since the discovery of fire. Much of the progress of human society has been made possible through access to more and more concentrated sources of energy. Early humans used wood and dung for cooking and warmth. The discovery of coal provided a more concentrated source of energy. Oil was another highly concentrated source.

Coal can be considered to be mostly pure carbon. The problem of acid rain is due to small sulfur impurities in coal. For the present, we are interested only in the carbon. Petroleum consists of a wide range of hydrocarbon molecules that contain only carbon and hydrogen. There are also a number of liquid fuels available for transportation that include oxygen atoms in the hydrocarbon molecules. The next section will give you the tools to understand the names and structures of these molecules.

LARGER MOLECULES OF CARBON, OXYGEN, AND HYDROGEN

One of the unique things about atoms of carbon is their ability to link with other atoms of carbon to form large structures containing hundreds or even thousands of carbons linked together. No other atom has this property to the same extent as carbon.

The chemistry of molecules containing carbon is called *organic chemistry*. This name derives from the fact that early chemists believed that organic molecules contained some *vital force* and could only be made by living organisms, hence the name organic chemistry. Great was the shock in the early 1800s when scientists made the first organic molecule from *inorganic* chemical starting materials. The history of this discovery is an interesting example of how discoveries in science can have an impact on the thinking of society and even on religion.

You can construct a large number of molecular models of organic chemicals with a model set. The easiest way to make them is to link the carbons together to form long chains and add hydrogens along the edges of the chain. One can make a series of these, each of which has one additional carbon atom in the chain with two additional hydrogens attached. The name for this type of molecule is *normal hydrocarbons*. You may also attach carbons along the side of the chain at any carbon to produce a branching effect. This class of molecules is known as *branched-chain hydrocarbons*.

The names of the normal hydrocarbons having one to ten carbon atoms are shown in Table 3. These names are important in that they form the root names of most organic molecules. The names for C-1 to C-4 hydrocarbons make sense only in that they reflect the discovery of these substances. Butane, for example, gets its name from the fact that butter contains a four-carbon compound. After C-4, the names derive from the Latin for the number. You will see that most of these names are familiar from the names of geometrical objects like a hexagon or octagon.

As the number of carbons becomes larger, the number of possible ways to construct molecules from them goes up very rapidly. Although there are only three C_5 hydrocarbons, there are 18 C_8 hydrocarbons and 75 ways to construct a C_{10} hydrocarbon. Different structures for the same molecular formula are called *isomers*. Gasoline

TABLE 3 Names of normal hydrocarbons

Number of Carbons	Normal Hydrocarbon	Root Name When Naming Functional Groups	Name of Group If Substituted on a Molecule
1	Methane	Meth	Methyl
2	Ethane	Eth	Ethyl
3	Propane	Prop	Propyl
4	Butane	But	Butyl
5	Pentane	Pent	Pentyl
6	Hexane	Hex	Hexyl
7	Heptane	Hept	Heptyl
8	Octane	Oct	Octyl
9	Nonane	Non	Nonyl
10	Decane	Dec	Decyl

consists primarily of isomers with eight carbons, and diesel fuel consists primarily of isomers of ten carbons.

Petroleum products are examples of hydrocarbons such as this. The fuel for much of our standard of living over the past century has been hydrocarbons. Hydrocarbon fuels have very high energy density (energy per gram or pound), and this makes them very useful as fuels. They are also valuable as starting materials for drugs, plastics, and other products. Everyone agrees that the crude oil will eventually run out, and that we must address this issue soon. It is also clear that if we were to burn all known petroleum reserves on Earth, the amount of CO_2 in the atmosphere would become dangerously high.

Naming Hydrocarbons

Hydrocarbons are named in a series of steps.

1. Find the longest chain. This can be more complicated than it seems because for branched hydrocarbons you need to check all possible paths from end to end to find the longest one. The longest chain of carbon atoms determines the root name for the molecule.
2. Number each carbon in the longest chain. Typically you should begin numbering from the end with the most nonhydrogen atoms attached.
3. Identify each group substituted on the chain by its substituent name and the number of the carbon where it is attached.

Example
Name the hydrocarbons in Figure 20.

Answers
Model A
The longest chain is six carbons, so the root name is hexane.
Two one-carbon (methyl) groups are attached at carbons number 3 and 4.
A possible name would be 3-methyl-4-methyl-hexane.

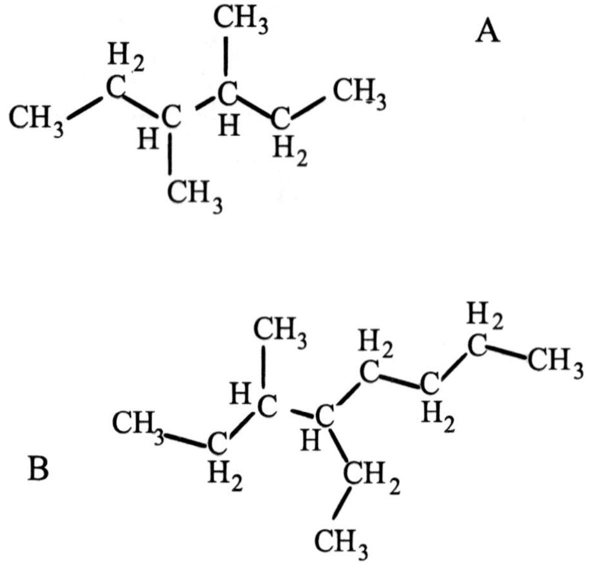

FIGURE 20 Example of hydrocarbon structures for naming.

When multiple substituents are present, the number of each is specified using prefixes di, tri, quadra, penta, and so on.

The final name is 3,4-dimethyl-hexane.

Model B
The longest chain is eight carbons, so the root name is octane.
A methyl group is attached to carbon number 3, and an ethyl group to carbon number 4.
The final name is 4-ethyl-3-methyl-octane.

Note that it also is possible to number these models from the other end. Model A would give the same name, but Model B would be named 5-ethyl-6-methyl-octane. One also needs to decide whether to list the ethyl group first or the methyl group. Scientists have decided to start numbering molecules like this from the end that gives the lowest numbers and to arrange substituent groups alphabetically.

ORGANIC MOLECULES CONTAINING OXYGEN

When we add oxygen atoms to molecules, either by using models or drawing structures on paper, there are four ways to attach the oxygens to the structure. An oxygen atom can be attached to a carbon and a hydrogen to give an—OH group attached to the chain. An oxygen can be attached between two carbons to give a C-O-C structure. An oxygen can be attached with a double bond to give a C=O structure. Finally, two oxygens can be attached together to give a structure C-O-O-H or C-O-O-C. Of course, molecules can contain any combination of these four schemes, and this leads to a large number of possible molecules.

All—OH groups may be referred to as *hydroxyl groups*. When the carbon to which the hydroxyl group (-OH) is attached has only carbon and hydrogen attached to its other three bonding sites, the structure is called an *alcohol*. This is shown in Figure 21. R is used in these structures to show that either C or H can be attached.

The term alcohol denotes one of the possible arrangements, and each unique arrangement is called a *functional group*. Hydroxyl groups that occur on carbons with other oxygens attached have other functional group names.

The names of simple alcohols are formed by writing the name of the hydrocarbon and replacing the -ane ending with -ol. The two-carbon alcohol is ethanol. For longer hydrocarbons, you indicate which carbon the hydroxyl group is attached to with the number.

The functional groups containing C=O include aldehydes, ketones, carboxylic acids, and esters. They are not common in fossil fuels.

Exercise
1. Draw structures for 1-propanol, 2-propanol, and 3-methyl-2-pentanol.

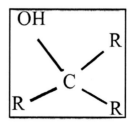

FIGURE 21 Structure of the alcohol group.

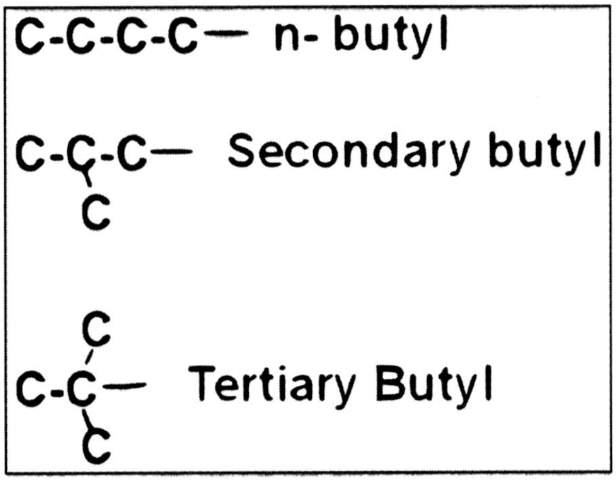

```
C-C-C-C—  n-butyl

C-C-C—  Secondary butyl
  |
  C

    C
    |
C-C-C—  Tertiary Butyl
    |
    C
```

FIGURE 22 Possible structures of a four-carbon molecule.

The other single-bonded oxygen functional group C–O–C is called the *ether* functional group. Ethers are named by using the names of the two substituent groups on the oxygen and then adding the word "ether." Diethyl ether $CH_3CH_2OCH_2CH_3$ has been used as an anesthetic for over a century. It is also highly flammable and dangerous to work with.

2. Four carbon hydrocarbons have three possible isomers as shown in Figure 22. Draw a structure for methyl tertiary butyl ether. This is a molecule that was used for a decade or more as a gasoline additive to replace tetraethyl lead.

3. Draw six isomers of C_8H_{18}.

Appendix 3. Quantitative Look at Combustion Reactions

When organic chemicals containing only the elements carbon, oxygen, and hydrogen react *completely* with oxygen, there are only two products formed: carbon dioxide (CO_2) and water (H_2O). If you have access to a model kit, build a model for ethane (C_2H_6). Also, take eight oxygen atoms from your model kit and connect them together with bonds to represent the oxygen molecules. The reaction of ethane with oxygen is described by the general equation:

$$C_2H_6 + O_2 \rightarrow CO_2 + H_2O$$

Take the model of ethane and the oxygen molecules. Take the ethane apart, and use the parts to construct CO_2 and H_2O. Now count the number of molecules of products you have made. You should have two molecules of carbon dioxide and three molecules of water. We write the number of each type of atom or molecule to the left of the formula to indicate how many are used in the reaction. These are called the coefficients of the reactants and products.

$$C_2H_6 + 4O_2 \rightarrow 2CO_2 + 3H_2O + O$$

The problem in the reaction we have just created using our model is that only seven oxygen atoms are needed by the reaction. Because oxygen exists as O_2 molecules, this means that we left a half-molecule of oxygen behind. We would say this reaction is unbalanced because we have some spare atoms that we did not use. This occurs because oxygen always exists as the O_2 molecule. We can solve this problem by using a second molecule of ethane. Now the reaction will look like this:

$$2C_2H_6 + 7O_2 \rightarrow 4CO_2 + 6H_2O$$

When we write the equation for any chemical reaction, we must do so in such a way that all of the molecules on each side come out even. Fractions are not normally allowed. Using models to study this reaction is particularly informative because it make it clear that chemical reactions occur on discrete molecules that we can count and must keep track of.

To write a balanced chemical equation for a combustion reaction, follow these three steps:

1. Add a coefficient to the CO_2 equal to the total number of carbon atoms in the hydrocarbon.
2. Add a coefficient to the H_2O equal to half of the total hydrogens in the hydrocarbon.
3. Count the total oxygen atoms on the right side of the equation. Subtract any oxygen atoms in the hydrocarbon.
   a. If the resulting number is even, divide by 2 and use this as the coefficient of the O_2.
   b. If the number is odd, double the coefficients on all molecules and use the number as the coefficient for the O_2.

These three simple steps will allow you to balance any combustion reaction for compounds of carbon, hydrogen, and oxygen. Because we cannot use half-molecules of oxygen, the third step is designed to ensure that the oxygens come out even.

Examples
Example 1: Balancing the Reaction for the Combustion of Decane

Write the reactants and products:

$$C_{10}H_{22} + O_2 \rightarrow CO_2 + H_2O$$

Balance the carbon:

$$C_{10}H_{22} + O_2 \rightarrow 10CO_2 + H_2O$$

Balance hydrogen:

$$C_{10}H_{22} + O_2 \rightarrow 10CO_2 + 11H_2O$$

Balance the oxygen. First count the oxygens on the right side—31 is an odd number. So double the coefficients of all atoms except O_2:

$$2C_{10}H_{22} + O_2 \rightarrow 20CO_2 + 22H_2O$$

Add the coefficient 31 to the oxygen:

$$2C_{10}H_{22} + 31O_2 \rightarrow 20CO_2 + 22H_2O$$

Now check it by adding the atoms on each side:

Carbons = 20
Hydrogens = 44
Oxygens = 62

Example 2: Balancing the Reaction for the Oxidation of C_3H_8O

Write the reactants and products:

$$C_3H_8O + O_2 \rightarrow CO_2 + H_2O$$

Balance the carbon:

$$C_3H_8O + O_2 \rightarrow 3CO_2 + H_2O$$

Balance the hydrogen:

$$C_3H_8O + O_2 \rightarrow 3CO_2 + 4H_2O$$

Balance the oxygen. Count the oxygens on the right—10. Subtract one oxygen from C_3H_8O to get 9, an odd number.
Double the coefficients:

$$2C_3H_8O + O_2 \rightarrow 6CO_2 + 8H_2O$$

Add the coefficient to oxygen:

$$2C_3H_8O + 9O_2 \rightarrow 6CO_2 + 8H_2O$$

Exercises

1. Write a descriptive essay in which you detail the exact steps you followed in the process of converting the models of ethane and oxygen into the models of carbon dioxide and water. Speculate on whether the real reaction might involve similar steps.
2. Balance the reaction for the combustion of C_6H_{14}.
3. Balance the reaction for the combustion of C_5H_{12}.
4. Balance the reaction for the combustion of $C_2H_4O_2$.

STOICHIOMETRY CALCULATIONS AND WEIGHTS OF REACTANTS AND PRODUCTS

Chemical reactions represent recipes for processes involving discrete atoms and molecules. The problem faced by chemists is that individual atoms weigh so little that we cannot possibly weigh them. An atom of carbon weighs 0.0000000000000000000000199 grams. That is 1.99×10^{-19} g if we write it in scientific notation. The solution to this problem was first proposed by the Italian scientist Amedeo Avogadro in the early nineteenth century and led to one of the most useful concepts in chemistry, the mole.

Counting by Weighing

If you wanted to count the number of M&Ms in a large bag, you could sit down and actually count all of them. Alternatively, you could accurately weigh a small number of M&Ms and determine the average weight per M&M. Then by weighing the large bag, you could determine the number of M&Ms present by dividing the total weight by the weight of an average M&M. Counting by weighing is not a foreign concept. Banks use it to count coins, and it is also used to count screws, nails, and other things in industry.

Exercise

5. Ten nails weighed on an accurate scale are 38.45 g. How many nails are in a box containing 238.39 g of nails?

We could apply the same concept to atoms by using the weight of a single atom and counting the number of atoms in any particular sample of a hydrocarbon. That would allow us to calculate the weight of reactants and products, but it would require the use of scientific notation throughout.

It is simpler to scale the entire process up to deal with a large collection of atoms or molecules that can be weighed in the laboratory. Again, this is a familiar concept in everyday life. Stores sell eggs by the dozen rather than one at a time. They sell cans of soda in groups of six or twenty-four. Beer is often sold in groups of six or twenty-four or even thirty cans. We are familiar with six-packs, cases, and possibly with reams of paper.

The number of particles required to move atoms and molecules up to a size we can see is very large. In fact, we have agreed to standardize the number as 6.023×10^{23}. This unit is called the *mole*. A mole of carbon is 6.023×10^{23} atoms of carbon. A mole of baseballs would be 6.023×10^{23} baseballs. This is a very large number indeed.

The mole is formally defined as 6.023×10^{23} of anything, and it is further defined as the number of carbon atoms in exactly 12 grams of pure carbon. Carbon is currently the standard element used to define the relative weight of all the other elements in the periodic table. The standard has changed over the years from hydrogen to oxygen, and now it is carbon. There will probably be further changes in the future as chemistry evolves to take advantage of new technology.

The reason for selecting this particular number as the collection to be used in our calculations is that the weight of a mole of any substance is the atomic or molecular weight in grams. (If we worked in ounces or pounds, we would have arrived at a different number to make this work.) The importance of this constant was first established by the scientist Avogadro, and the number bears his name, the Avogadro number. The good news is that you will probably never need to actually use this number. This value is sometimes used in computations, but it is unnecessary in the simple stoichiometry calculations we will do in this course.

What is essential is to realize that we can read chemical equations either in terms of molecules or moles. When we read them in terms of moles, we are thinking in terms of the number of particles, but we can do the calculations in grams.

Applying the Mole Concept

Let us now examine the amount of carbon dioxide and water produced in the combustion of any organic chemical. Because many organic chemicals are used as fuels, either as their primary use or during waste disposal, this calculation provides a way to determine which fuel will give the most of each product. One reason to do these calculations is to evaluate the amount of CO_2 emitted when various fuels are burned. By the end of this section you will be able to determine both the weight of each product and the amount of energy produced.

In the next example, we will step through the process of determining how much CO_2 is produced when 10 g of three fuels—natural gas, propane, and gasoline—are used as fuel.

Example

Calculate the weight of carbon dioxide produced by combustion of 10 g of each of the following fuels:

Methane (natural gas)
Propane (bottled gas)
Octane (gasoline)

Step 1: Determining the balanced reaction for each combustion reaction

$2CH_4 + 4O_2 \rightarrow 2CO_2 + 4H_2O$

$C_3H_8 + 5O_2 \rightarrow 3CO_2 + 4H_2O$

$2C_8H_{18} + 25O_2 \rightarrow 16CO_2 + 18H_2O$

The balanced chemical reaction is a recipe, just like the recipe for ham sandwiches or s'mores. It tells how *many* of each ingredient is used and how many products are formed. The next step is to use the mole concept to convert the known mass (10 g) of each fuel to the number of moles (abbreviated as "mol"). This is counting by weighing in action. We determine the weight of a mole of each substance by adding the weight of all atoms present.

Step 2: Determining the molecular weight of each substance
$CH_4 = 12$ g/mol $\times 1$ C $+ 1$ g/mol $\times 4$ H $= 16$ g/mol

$C_3H_8 = 3$ C $\times 12$ g/mol $+ 8$ H $\times 1$ g/mol $= 44$ g/mol

$C_8H_{18} = 8$ C $\times 12$ g/mol $+ 18$ H $\times 1$ g/mol $= 114$ g/mol

$O_2 = 2$ O $\times 16$ g/mol $= 32$ g/mol

$CO_2 = 1$ C $\times 12$ g/mol $+ 2$ O $\times 16$ g/mol $= 44$ g/mol

The number of moles is found by dividing the mass present by the molecular weight, just as we did in the M&Ms example. The way we prefer to show this is as a multiplication with the molecular weight in the denominator of a fraction. The reason we set the problem up this way is so that we can examine the units and confirm we are doing it correctly.

Step 3: Determining the number of moles of each substance in 10 g
10 g $CH_4 \times 1$ mol $CH_4/16$ g $CH_4 = 0.625$ mol CH_4

10 g $C_3H_8 \times 1$ mol $C_3H_8/44$ g $C_3H_8 = 0.227$ mol C_3H_8

10 g $C_8H_{18} \times 1$ mol $C_8H_{18}/114$ g $C_8H_{18} = 0.0877$ mol C_8H_{18}

Note that in the step above, the units of g in the mass cancel the mass in the denominator of the fraction leaving moles as the unit for the answer.

The next step is to use the recipe and the number of moles of reactants to determine the moles of product. You know how to do this intuitively if you have ever had to figure out how many ham sandwiches or s'mores you can make from a collection of food you have purchased.

For example, if the recipe for s'mores is

1 chocolate bar + 2 graham crackers + 2 marshmallow → 2 s'more

then if you have three chocolate bars, you know intuitively that you can make six s'mores, provided that you have enough graham crackers and marshmallows (and we always seem to have plenty of everything but the chocolate bars). The calculation you are doing in your head could be written as

3 chocolate bars $\times$ 2 s'mores/1 chocolate bar = 6 s'mores

Note that we use the units and the ratios of reactants and products from the recipe to confirm we are doing the problem correctly.

Step 4: Determining the number of moles of carbon dioxide produced from the stoichiometry of the balanced reaction
0.625 mol $CH_4 \times 2$ mol $CO_2/2$ mol $CH_4 = 0.625$ mol CO_2

0.227 mol $C_3H_8 \times 3$ mol $CO_2/1$ mol $C_3H_8 = 0.681$ mol CO_2

0.0877 mol $C_8H_{18} \times 18$ mol $CO_2/2$ mol $C_8H_{18} = 0.789$ mol CO_2

Again, the moles of the reactant cancel, and the answer has units of moles for CO_2.

Finally, we can convert the moles of product to grams using the molecular weight of product.

Step 5: Converting the number of moles of carbon dioxide to the weight in grams
0.625 mol $CO_2 \times 44$ g $CO_2/1$ mol $CO_2 = 27.5$ g CO_2

0.681 mol $CO_2 \times 44$ g $CO_2/1$ mol $CO_2 = 30.0$ g CO_2

0.789 mol $CO_2 \times 44$ g $CO_2/1$ mol $CO_2 = 34.7$ g CO_2

It is possible to write any of these calculations as a single long series of multiplications:

10 g × 1 mol CH_4/16 g × 2 mol CO_2/2 mol CH_4 × 44 g CO_2/1 mol CO_2 = 27.5 g CO_2

To summarize the process:

1. Balance the chemical reaction.
2. Convert mass to moles.
3. Use the stoichiometry ratio in the balanced reaction to get moles of products from moles of reactant.
4. Convert moles back to grams.

Exercises

6. Several fuels containing oxygen are now being used commercially. Calculate the amount of carbon dioxide produced by the combustion of 10 g each of the following:

   Methanol.
   Ethanol.
   Methyl-butyl ether.

7. Calculate the weight of carbon dioxide produced by the combustion of one gallon of octane (1 gal of octane weighs 2,620 g). Use the example in the text to provide data for this problem.

ENERGY ASSOCIATED WITH COMBUSTION REACTIONS

We all know that flames give off heat. This energy associated with a combustion reaction is also a part of the reaction. We can show it by writing the reaction in the following manner:

$CH_4 + 2O_2 \leftrightarrow CO_2 + 2H_2O + $ Heat

The amount of heat associated with a combustion reaction is, of course, a very important feature of these reactions. We depend on this reaction product to heat our homes and to power our cars.

It is possible to deal with the heat in a quantitative manner just as we deal with the number of molecules involved in the reaction. Chemists have found that *the energy content of each type of bond is relatively constant.*

When we did the model exercise on the combustion of ethane, we learned that the first step was to take apart the models of the reactants. This meant breaking all the C–C, C–H, and O=O bonds to get the free atoms. Breaking chemical bonds requires energy. That is why you need a match to light a fire. The match breaks the fuel into atoms that can then form new bonds to make the products. When chemical bonds form, energy is released, usually as heat. Therefore, it becomes possible to predict approximately how much energy is released or taken up in a chemical reaction by considering all the bonds broken and formed.

Calorimetry

It is possible to measure the amount of energy absorbed or released in a chemical reaction very accurately. The process is called *calorimetry*. The heat released when organic chemicals are burned is measured by burning the compounds in pure oxygen in a closed system. Because a large amount of energy is released, the container must be very strong; the apparatus for this type of calorimetry is called a *bomb calorimeter* and is shown in the diagram in Figure 23. Occasionally these actually do explode.

CALCULATIONS USING BOND ENERGIES

The numerical values of the energy changes involved in breaking and forming bonds during combustion are shown in Table 4. Some additional bond energies relevant to environmental reactions are also included. You can use these values to approximate whether the reaction will require a net input of energy or will produce energy as a product. In combustion reactions, we are only interested in the formation of bonds between carbon and oxygen and between oxygen and hydrogen. We can have high confidence that these will be constant.

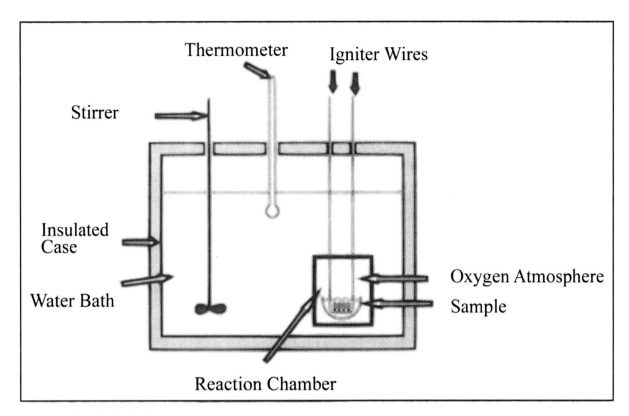

FIGURE 23 Bomb calorimeter.

The units of the energies in the table are kilojoules (kJ) per mole. This term probably means very little to you. To put it into prospective, 418 kJ is enough energy to heat a liter of water (about a quart) from the temperature of ice to the boiling point. This means that you could heat the water for four cups of coffee with this much energy. Thus, 100 kJ is about enough energy to make a cup of coffee from an ice water start.

We can use the information in Table 4 to calculate the approximate energy change for a combustion reaction. Let us consider the combustion of ethane that we carried out in the model exercise.

One molecule of ethane (CH_3-CH_3) has the following bonds:

1 C-C

6 C-H

Therefore one mole of ethane has 1 mole of C-C bonds and 6 moles of C-H bonds. Remember the term mole simply means we have 6.0×10^{23} of something.

We calculate the energy required to disassemble the molecule by multiplying the energy per bond by the total number of the bonds:

1 C-C × 347 kJ/ mol C-C + 6 C-H × 413 kJ/ mol C-H = 2,825 kJ/mol ethane

The balanced reaction requires two molecules of ethane, so we must double the above number to get the total energy required for the two molecules of ethane in the balanced reaction.

2,825 kJ/mol ethane × 2 mol ethane = 5,650 kJ

TABLE 4 Energies associated with specific chemical bonds

Bond	Energy kJ/mol	Bond	Energy kJ/mol	Bond	Energy kJ/mol	Bond	Energy kJ/mol
C-H	413	O=O	495	F-F	154	N-O	200
C-C	347	O-H	467	Cl-Cl	239	N=O	605
C=C	614	H-H	432	H-F	565	N=O	607
C≡C	839	C-O	358	H-Cl	428	N-N	165
C≡O	1071	C=O (carbonyl)	745	S=O	515	N=N	420
C-C in coal (one C-C per carbon)	709	C=O (CO_2 only)	799			N≡N	943

We also must disassemble seven molecules of oxygen. The energy required is seven times the energy of a single O=O, or

7 O=O × 495 kJ/mol O=O = 3,465 kJ

The total energy required to take apart all of the reactants in the balanced reaction is then the total for the two molecules of ethane and the seven molecules of oxygen, or 9,115 kJ.

Now we can calculate the total energy we get back when we form the bonds between carbon and oxygen in carbon dioxide and between oxygen and hydrogen in water. Again, we count the total number of bonds of each type, then we multiply by the energy of each type of bond.

4 CO_2 = 8 C=O bonds 8 mol C=O × 799 kJ/mol C=O = 6,392 kJ

6 H_2O = 12 O-H bonds 12 mol O-H × 467 kJ/mol O-H = 5,604 kJ

The total energy return from forming the bonds is 11,996 kJ. This is obviously more than the energy we had to invest to take apart the atoms originally. The difference between the energy required to break up the reactant molecules and that produced by forming the product molecules is the total energy produced in the reaction. In this case, 11,996 kJ − 9,115 kJ = 2,881 kJ.

If we use the energy from 2 moles of ethane to boil water for coffee, we find that we can make about 28 cups of coffee: 2 moles of ethane in the liquid state would fill about six tablespoons. To get the energy/mole of ethane, you must take account of the fact that this calculation involves two moles of ethane:

2,881 kJ/2 mol × ½ = 1,440.5 kJ/mol × 30 g/w mol = 48.1 kJ/g ethane

You can use this approach to calculate the approximate energy produced by any of the combustion reactions we are examining. We can summarize the process in a simple equation:

Total energy = SUM(All bonds broken) − SUM (All bonds formed)

The equation will give a negative total energy for combustion reactions. For reasons of definition that do not really matter here, negative energy means that the heat is given off. This is a bit like a negative change in altitude on the ski slope showing you are going down and picking up speed. If you calculate the total energy change for a reaction and get a positive value, this means the reaction uses up heat.

Exercises

8. There are four fossil fuels in common use in transportation. These include octane (gasoline), methanol, ethanol, and methane.
   a. Write the reactions for the complete combustion of each of fuels. (They are already balanced in the example in the text and in Exercise 6.)
   b. Calculate the heat produced when 1 mole of each is burned.
   c. Calculate the amount of heat produced when 100 g of each of these fuels is burned.
9. Concern about the greenhouse effect has focused attention on the amount of carbon dioxide produced by various fuels. Calculate the grams of carbon dioxide produced per kilojoule of energy by the combustion of each fuel in Exercise 8. The calculation is done per gram rather than per mole because you have to carry the fuel with you and the weight of the fuel is important. Compare methane, methanol, ethanol, and octane. Which fuel would you select if the greenhouse effect were the only consideration?
10. Calculate the heat produced from the combustion of coal (pure carbon) per mole and per 100 g. Then calculate the carbon dioxide produced per 100 g. Finally, determine the carbon dioxide produced per kilojoule of energy from burning coal. Compare coal with the other fuels you analyzed in Problem 9. What fuel is best for generating electricity from the standpoint of climate change.
11. What factors are important in selecting a fuel for cars? Discuss the various factors that must be considered and their relative importance.

Appendix 4. Leaked Draft Document by Danish Delegates

This document has been edited to remove the sections not relevant to the game. Important points in the debate are in **bold**.

DRAFT 271109
Decision 1/CP.15
(Decision 1/CMP.5 in separate document)
Adoption of
The Copenhagen Agreement
Under the United Nations Framework Convention on Climate Change

The Conference of the Parties,
Pursuant to the Bali Road Map adopted by the Conference of the Parties at its thirteenth session,
Acknowledging and building on the work by the Ad Hoc Working Group on Long-Term Cooperative Action under the Convention and the Ad Hoc Working Group on Further Commitments for Annex I Parties under the Kyoto Protocol,
Sharing a commitment to take immediate and enhanced national action under the Convention in pursuit of its ultimate objective, and in accordance with its principles and commitments including the principle of common but differentiated responsibilities and respective capabilities,
Seeking at the same time to move ahead promptly to take action related to address climate change,
Believing it imperative that the Parties continue to work together constructively to strengthen the world's ability to combat climate change,
Affirming the need to continue negotiations pursuant to decisions taken at COP13 and COP15, with a view to agreeing on **a comprehensive legal framework under the Convention no later than 2015**

Decides to adopt this political agreement (hereinafter "the Copenhagen Agreement"), which will become effective immediately.

THE COPENHAGEN AGREEMENT
1. The Parties to the United Nations Framework Convention on Climate Change (hereinafter "the Parties") seek to further the implementation of the Convention in a manner that pursues its ultimate objective as stated in its Article 2, that recalls its provisions, and that is guided by the principles in Article 3.

I. A Shared Vision for Long-Term Cooperative Action
2. The Parties underline that climate change is one of the greatest challenges of our time and commit to a vigorous response through immediate ambitious national action and strengthened international cooperation with a view to limit global average temperature rise to a maximum of **2 degrees** above pre-industrial levels. The Parties are convinced of the need to address climate change bearing in mind that social and economic development and poverty eradication are the first and overriding priorities in developing countries. The Parties note that the largest share of historical global emissions of greenhouse gases originates in developed countries, and that per capita emissions in many developing countries are still relatively low.
The Parties recognize the urgency of addressing the need for enhanced action on adaptation to climate change. They are equally convinced that moving to a low-emission economy is an opportunity to promote continued economic growth and sustainable development in all countries recogniz-

ing that gender equality is essential in achieving sustainable development. In this regard, the Parties:

- Commit to take action to mitigate climate change based on their common but differentiated responsibilities and respective capabilities,
- Commit to take action on adaptation including international support assisting the poorest and most vulnerable countries,
- Commit to strengthen the international architecture for the provision of substantially increased finance for climate efforts in developing countries,
- Commit to establish a technology mechanism to promote the development, transfer and deployment of environmentally sustainable technologies in support of mitigation and adaptation efforts.

[. . .]

3. Recalling the ultimate objective of the Convention, the Parties stress the urgency of action on both mitigation and adaptation and recognize the scientific view that the increase in global average temperature above pre-industrial levels ought not to exceed **2 degrees C**. In this regard, the Parties:

- Support the goal of a peak of global emissions as soon as possible, but no later than **[2020]**, acknowledging that developed countries collectively have peaked and that the timeframe for peaking will be longer in developing countries,
- Support the goal of a reduction of global annual emissions in 2050 by at least 50 percent versus 1990 annual emissions, equivalent to at least 58 percent versus 2005 annual emissions. The Parties contributions towards the goal should take into account common but different responsibility and respective capabilities and a long term convergence of per capita emissions.

[*Note:* Section I commits to a 2 degree limit to temperature increase and recognizes that different countries have different needs and responsibilities. It set 2020 as the year after which global emissions will begin to decline and sets a 50 percent reduction goal in 2050 relative to 1990 emissions. It also calls for the per capita emissions of all countries to "converge," meaning that no country should have the right to higher per capita emissions than another.]

II. Adaptation

4. The adverse effects of climate change are already taking place and are posing a serious threat to the social and economic development of all countries. This is particularly true in the most vulnerable developing countries, which will be disproportionally affected. The adverse impact of climate change will constitute an additional burden on developing countries' efforts to reduce poverty, to attain sustainable development and to achieve the United Nations Millennium Development Goals. Both adaptation and mitigation efforts are fundamental to the fight against climate change. Adaptation must include action to reduce risk and vulnerability, taking into account gender equality, and build resilience in order to reduce the threats, loss and damages to livelihoods and ecosystems from disasters caused by extreme weather events and from slow-onset events caused by gradual climate change. Recognizing that the impact of climate change will differ according to regional and national circumstances, planning and implementation of adaptation actions must be considered in the context of the social, economic and environmental policies of each country. Adaptation action at national level will be a country driven process taking into account national development priorities and plans.

5. In this regard, the Parties endorse the adaptation framework in decision X4/CP.15 with the objective of reducing vulnerability and building resilience to present and future effects of climate change through national action and international cooperation. This

includes the provision of finance, technology and capacity building in the immediate, medium and long term. Support should be provided with priority for the poorest and most vulnerable countries. In the context of this Framework institutional arrangements will be established over time to support Parties' actions and provide technical assistance including for risk reduction and provide financial risk transfer such as insurance. Further, this will include a system to ensure mutual accountability with monitoring, review and assessment of support and actions and share lessons learned. A share of fast-start financing comprising **[$X]** for 2010–12 will be provided through existing channels, including the Adaptation Fund, to implement actions identified in National Adaptation Programmes of Action and other urgent needs and to build capacity for further planning.

[*Note:* Section II notes the special needs of developing countries for both adaptation to the changing climate and mitigation of the damage caused by the changes.]

III. Mitigation

6. The shared vision limiting global average temperature rise to a maximum of 2 degrees above pre-industrial levels is addressed by nationally appropriate mitigation contributions to be carried out by the Parties consistent with the principle of common but differentiated responsibilities and respective capabilities and with developed countries taking the lead.

Developed Countries nationally appropriate mitigation commitments and actions

7. The developed country Parties commit to individual national economy wide targets for 2020. The targets in Attachment A would expect to yield aggregate emissions reductions by **X1 percent by 2020 versus 2000**. The purchase of international offset **credits will play a supplementary role to domestic action. The developed country Parties support a goal to reduce their emissions of greenhouse gases in aggregate by 80% or more by 2050 versus 1990 (X3 percent versus 2005).**

8. Attachment A reflects the individual economy-wide targets, including quantified emission limitation and reduction objectives by all the developed country Parties.

Developing countries nationally appropriate mitigation actions

[. . .]

10. Attachment B reflects individual commitments to nationally appropriate mitigation actions by developing country Parties. Developing country parties which have not reflected their contributions at COP15 should do so before **2015**, except least developed countries. A developing country Party may subsequently amend its national contribution to register additional national appropriate mitigation actions which increase its overall mitigation outcome.

11. **A Registry in the form of a database under UNFCCC is established in order to enable the international recognition of developing country mitigation action.** The Registry shall include supported mitigation actions that meet agreed MRV specifications and unsupported actions that are subject to national MRV based on internationally agreed guidelines and a consultative review under UNFCCC. Developing countries commit to inscribe supported nationally appropriate mitigation actions in the Registry and indicate the expected emissions outcomes. Unsupported action shall, except for the least developed countries which may do so at their own discretion, be inscribed via the National Communications and can be inscribed directly in the Registry beforehand on a voluntary basis.

Reducing Emissions from Deforestation and Forest Degradation

12. Reducing emissions from deforestation and forest degradation is an important aspect of the necessary response to climate change. **Developing countries should contribute to enhanced**

mitigation actions through reducing emissions from deforestation and forest degradation, maintaining existing and enhancing carbon stocks, and enhancing removals by increasing forest cover. Parties underline the importance of enhanced and sustained financial resources and positive incentives for developing countries to, through a series of phases, build capacity and undertake actions that result in measurable, reportable and verifiable greenhouse gas emission reductions and removal and changes in forest carbon stocks in relation to reference emission levels.
[...]

National Policies

15. The Parties commit to further integrate low-emission development policies into national planning. The Parties commit to rationalize and **phase out over the medium term inefficient fossil fuel subsidies that encourage wasteful consumption.** As we do that, we recognize the importance of providing those in need with the ability to purchase essential energy services, including through the use of targeted cash transfers and other appropriate mechanisms. In addition, the Parties commit to work towards adopting domestic policies aiming at payment for actual consumption of energy. Furthermore, transparency concerning consumption and cost of energy should be increased.
[...]

V. Financial resources and investments to support actions on mitigation, adaptation, capacity-building and technology cooperation

19. Substantially scaled up financial resources will be needed to address mitigation, adaptation, technology and capacity building. It is essential to strengthen the international financial architecture for assisting the developing countries in dealing with climate change and to improve access to financial support. Resources will derive from multiple sources and flow through multiple bilateral and multilateral channels.

20. The Parties share the view that the strengthened financial architecture should be able to handle gradually scaled up international public support. International public finance support to developing countries **[should/shall]** reach the order of **[X] billion** USD in 2020 on the basis of appropriate increases in mitigation and adaptation efforts by developing countries.

21. **The Parties** *confirm* **climate financing committed under this agreement as new and additional resources that supplement existing international public financial flows otherwise available for developing countries in support of poverty alleviation and the continued progress towards the Millennium Development Goals.** In this regard:

- Developed country parties commit to deliver upfront public financing for 2010–201[2] corresponding on average to **[10] billion** USD annually for early action, capacity building, technology and strengthening adaptation and mitigation readiness in developing countries as set forth in Attachment C;
- From [2013] The Parties commit to regularly review appropriateness of contributions and the circle of contributors against indicators of fairness based on GDP and emissions levels and taking into account the level of development as set forth in Attachment C.

22. Recalling article 4 of the Convention, Parties decide that a Climate Fund be established as an operating entity of the Financial Mechanism of the Convention, which should function under the guidance of and be accountable to the COP as set forth in article 11 of the Convention. The Fund should be operated by a board with balanced representation, which will develop the operational guidelines for the Fund and decide on specific allocation to programmes and projects. The COP will formally elect members of the Fund Board and endorse the operational guidelines and modalities for the Fund. The Fund should complement and

maximise global efforts to fight climate change through up-scaled support for climate efforts in the developing countries, including mitigation, adaptation, technology and capacity-building. Support from the Fund may be channeled through multilateral institutions or directly to national entities based on agreed criteria. Parties commit to allocate an initial amount of **[$x]** to the Fund as part of their international public climate support. Medium term funding should be based on a share of no less than **[y%]** of the overall international public support. Parties decide to operationalise the work of the Fund following the modalities set forth in annex/decision [Y].

23. In the context of the commitment in paragraph [14] Parties commit to global financing contributions from international aviation and international maritime transport generated through instruments developed and implemented by the ICAO and IMO respectively should be channeled through the Climate Fund from [2013], [mainly for adaption purposes], taking into account the principle of common but differentiated responsibility.

24. **To enhance transparency and overview** The Parties decide to establish an International Climate Financing Board under the UNFCCC to monitor and review international financing for climate action and in this context identify any gaps and imbalances in the international financing for mitigation and adaptation actions that may arise. **The Board will consist of [x] representatives from developed countries and [y] representatives from developing countries. [Z] Representatives from international institutions will participate in the Board as permanent observers. Decision making will be by consensus. [If all efforts to reach a compromise have been exhausted and no agreement has been reached, decisions shall be taken by a two-thirds majority]. The UNFCCC Secretariat will serve as secretariat for the International Climate Financing Board. Parties endorse the further guidelines as set out in attachment D and** decision X7/CP.15.

[. . .]

VI. Measurement, Reporting and Verification and Improved National Communications

26. The Parties commit to robust measurement, reporting and verification (MRV) of the commitments undertaken in this Agreement and to review global progress in addressing climate change. The Parties endorse the further guidelines as set out in decision X7/CP.15.

Measurement, Reporting and Verification for Developed Countries

27. In order to promote **transparency** and accountability the developed country Parties will report on the implementation of their individual mitigation commitments or actions in Annex A, including methodologies and assumptions used. The implementation of the respective mitigation contributions and the related reductions are subject to international measurement, reporting and verification and each developed country Party is to report on emission reductions achieved in relation to targets in Attachment A utilising a common methodology. Finance, technology and capacity building for developing countries actions are subject to robust MRV. Provision of international public climate financing should be verified in conjunction with the MRV of supported action and according to international guidelines.

Measurement, Reporting and Verification for Developing Countries

28. In order to promote transparency and accountability the developing country Parties will report on the implementation of their individual mitigation actions and emission outcomes achieved in relation to their estimates in Attachment B. The supported mitigation actions and the related reductions are subject to robust MRV. MRV of supported actions must verify that financing as well as action is delivering in full towards commitments. Implementation of developing country mitigation actions that are not externally supported will be subject to national MRV based on

international agreed guidelines and a consultative review under UNFCCC.

The Registry

29. Parties decide to establish a Registry that will be managed and operated independently by a professional secretariat which shall perform its tasks to the highest standards of professionalism and objectivity. The secretariat shall further prepare and propose the accounting standards for MRV of specific mitigation action and of financing. [further tasks]

Improved National Communications

30. Noting that low-emission development is indispensable to sustainable development and recognizing that development strategies and priorities are sovereign national decisions, the Parties decide to strengthen the reporting regime and to enhance the forward looking aspects in the National Communication by including mitigation plans. The Parties decide

- That Parties are to provide their greenhouse gas inventories on an annual basis with the exception that the developing countries can provide updates on a biannual basis and the least developed country Parties on a triennial basis;
- National Communications should be provided every 2 years. The Parties endorse the further guidelines as set out in decision X1/CP.15.
- To include a forward looking mitigation plan would help frame actions planned in the near- and medium- term (2020) in the context of longer-term goals (2050). The plans should describe countries' current mitigation and energy policy frameworks including regulation and pricing and mitigation potential. For developing countries, these plans will help facilitate access to support for mitigation actions anchored in the plans. The Parties, except the least developed countries which may contribute at their own discretion, are invited to put forward National Communication including forward looking plans as early as possible and [before 31 May 2011] in accordance with revised national communication principles and procedures in [Attachment X]/[Decision X/CP.15]. The plans will be updated every 2 year.
- The Parties will report, as applicable, on support received and support provided to developing countries for their actions in National Communications. A comprehensive set of statistics for climate change finance will be established enabling transparent monitoring of both provision of finance and supported climate actions. Financial flows from the international carbon market should be monitored and recognized separately.
- To enhance and expand the scope of the review of inventories and National Communications a consultation procedure is established under the SBI. The Parties undertake such consultations on the basis of input prepared by a newly established Expert group on National Communications that consider National Communications, including countries' forward looking plans.

VII. The Copenhagen Process

[*Note:* Because Copenhagen did not reach a legally binding treaty, this section defines the process to reach such a treaty.]

31. The Parties underline their commitment to immediate action pursuant to this Agreement. Furthermore, the Parties:

- Decide to continue negotiations pursuant to decisions taken at COP13 and COP15, with a view to agreeing on a comprehensive legal framework under the Convention no later than COPXX
- Decide on a review of commitments and actions under the Convention to be started in

2014 and completed in 2015 with a view of enhancing commitments and actions on mitigation and adaptation, and climate finance to achieve the Convention's ultimate objective and paragraph 2 and 3 above taking into account the conclusions of the Fifth Assessment Report of IPCC.
- Will keep track of overall efforts with a view to ensure full transparency and allow The Parties to strengthen their collective commitments and efforts if necessary to deliver in full towards addressing the climate challenge.

32. The Parties commit to work together in international organizations, including international financial institutions, to further integrate climate aspects in their activities, including country reviews.

Attachment A: National Mitigation Contributions

NAME OF COUNTRY *[developed countries]*
National Emissions Trajectory towards 2050
For example, reduce emissions to at least X per cent below X levels by 20XX, with milestones specified
National Contribution: Mitigation in 2020

Brief Description	Emissions outcomes, including baseline and timeframe
Economy wide quantified emission reduction target in the form of QELROs	*X per cent reduction on 1990/2005 levels by 2020*
Quantified emission limitation and reduction objective pursuant to or to be effected by domestic law and regulatory authority.	*X per cent reduction on 1990/2005 levels by 2025*
Use of carbon offsets	*A maximum of X per cent being achieved with REDD credits included*
	A maximum of Y per cent being achieved without REDD credits included
Policies and measures contributing to economy wide target if desired	*Contributing to economy wide target*
Eg Renewable energy target	
Eg Regulation on land clearing	

Attachment B: National Mitigation Contributions

NAME OF COUNTRY: *[developing country minus least developing countries]*
National Emissions Ambition

For example, specific milestones towards peaking and reverse emissions

National Contribution: Mitigation in 2020

Brief Description	Emissions Outcomes Expected, Including Baseline and Timeframe
Eg Forest reference level	*Eg X Mt CO_2 eq relative to a base year of XXXX by 20XX or defined business as usual of X Mt CO_2 eq*
Eg Renewable energy target	*Eg X Mt CO_2 eq relative to a base year of XXXX by 20XX or defined business as usual of X Mt CO_2 eq*
Eg Regulations on land clearing	*Eg X Mt CO_2 eq relative to a base year of XXXX by 20XX or defined business as usual of X Mt CO_2 eq*
Eg Emissions / carbon / energy intensity target	*Eg X hectares of clearing avoided.*
	Eg X percent relative to a base year of XXXX by 20XX or defined business as usual of X Mt CO_2 eq
	Eg X percent energy intensity improvement is implemented

Additional Mitigation Receiving Support

Brief Description	Emissions Outcomes Expected, Including Baseline and Timeframe	Total Costs (*all sources*)	International support (*component of total*)
Eg New Building Code: minimum energy efficiency for new constructions	*Eg X Mt CO_2 eq reduction relative to business as usual of X Mt CO_2 eq or base year of XXXX by 20XX*	*$X*	*$X*

LEAKED DRAFT DOCUMENT BY DANISH DELEGATES

Attachment C: Fast Start Finance

Country	National Contribution for 2010–2012
Listing of countries in alphabetical order	$X
Total pledge	[30]